PERILOUS PASSAGE

A Narrative of the Montana
Gold Rush, 1862–1863

This portrait of Edwin Ruthven Purple is from
In Memorium: Edwin R. Purple, *published*
by his brother Samuel S. Purple, M.D., after
Edwin died of pneumonia in 1879.

PERILOUS PASSAGE

A Narrative of the Montana
Gold Rush, 1862–1863

By Edwin Ruthven Purple
Edited by Kenneth N. Owens

Montana Historical Society Press
Helena

Cover design by Kathryn Fehlig
Typeset in Adobe Garamond
Printed by BookCrafters, Chelsea, Michigan

96 97 98 99 00 01 02 03 04   9 8 7 6 5 4 3 2

Library of Congress Cataloging-in-Publication Data

Purple, Edwin R. (Edwin Ruthven), 1831–1879.
    Perilous passage : a narrative of the Montana Gold Rush, 1862–1863 / by
  Edwin Ruthven Purple ; edited by Kenneth N. Owens.
        p. cm.
      Includes index.
      ISBN 0-917298-35-7 (hardback : alk. paper)
      ISBN 0917298-37-3 (pbk. : alk. paper)
        1. Montana—Gold discoveries. 2. Frontier and pioneer life—Montana.
  3. Purple, Edwin R. (Edwin Ruthven), 1831–1879. 4. Pioneers—Montana—
  Biography. I. Owens, Kenneth N.
  F731.P87 1995
  978.6' 02' 092—dc20
      [B]   95-33196
                                                                                   CIP

CONTENTS

ILLUSTRATIONS

ACKNOWLEDGMENTS

Preparation of Ed Purple's manuscript for publication has been an intriguing and enjoyable project. Among those who have helped along the way, I would like particularly to acknowledge the assistance of Margaret Heilbrun of the New-York Historical Society; Michael N. Landon of the Church Archives Division, Historical Department, Church of Jesus Christ of Latter-day Saints; Will Bagley of Salt Lake City; Robert Clark and Dorothea Simonson of the Montana Historical Society Library; Kathryn Otto and her staff at the Montana State Archives; and Lyndel Meikle of the Grant-Kohrs Ranch National Historic Site in Deer Lodge, Montana. I have a particular debt of gratitude to F. Lee Graves, Bannack City's historian, who led me on a personal tour to Montana's first "last best place," the scene of Ed Purple's last best adventures.

My discovery of Purple's literary legacy was an example of scholarly serendipity for which I owe thanks to my wife, Sally Owens. Some years ago she persuaded me to suspend my aversion to eastern locales and accompany her to New York City, where she was presenting a paper to the annual meeting of the American Society for Microbiology. After listening for awhile to world-famous scientists hold forth on arcane matters of which I was entirely ignorant, I went in search of more familiar environs at the New-York Historical Society. To find the Purple materials there was an unanticipated boon, offering its own reward as well as a fresh justification for the tax deductibility of the trip.

After preparing a first transcript of the narrative, I let the project languish until Charles Rankin, editor of the Montana Historical Society Press, expressed an interest. Chuck's encouragement and assistance, along with the work of his very able staff, has been crucial in bringing this account into print. I am also indebted to the Trustees of the Montana Historical Society Foundation for a research grant that assisted my efforts to retrace Ed Purple's 1862 travels and reclaim the documentary record related to his perilous passage among the other treasure seekers in the goldfields of early Montana.

Kenneth N. Owens
Sacramento, California

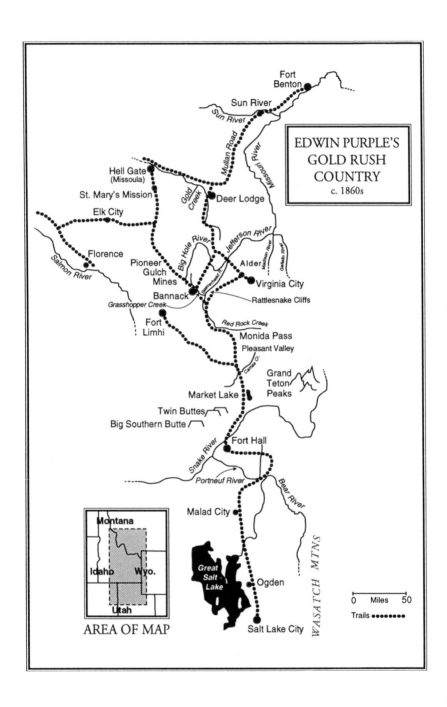

EDWIN PURPLE'S
GOLD RUSH
COUNTRY
c. 1860s

Fort Benton

Sun River

Sun River

Mullan Road

Missouri River

Hell Gate
(Missoula)

St. Mary's Mission

Elk City

Gold Creek

Deer Lodge

Jefferson River

Florence

Salmon River

Pioneer Gulch Mines

Big Hole River

Madison River

Gallatin River

Alder

Virginia City

Bannack

Beaverhead R.

Grasshopper Creek

Rattlesnake Cliffs

Fort Limhi

Red Rock Creek

Monida Pass

Pleasant Valley

Camas Cr.

Grand Teton Peaks

Market Lake

Twin Buttes

Big Southern Butte

Snake River

Fort Hall

Portneuf River

Bear River

Malad City

Montana

Idaho   Wyo.

Utah

AREA OF MAP

Great Salt Lake

Ogden

Salt Lake City

WASATCH MTNS

0   Miles   50

Trails ●●●●●●●●

x

INTRODUCTION

These pages contain an original narrative of the gold rush era in western Montana, although no one had yet applied the name "Montana" to this part of the northern Rockies at the time of the events described here. The author is Edwin Ruthven Purple, a native New Yorker who joined the California gold rush in 1850 and spent nearly a decade in an unsuccessful quest for wealth in the mining camps and boomtowns of northern California's Mother Lode region. In 1861, after a brief sojourn at Fort Yuma, California, Purple moved to Salt Lake City where he managed the local office of the Butterfield Overland Mail Company. The following spring reports reached Salt Lake City of new gold discoveries somewhere in the northern Rocky Mountains. Vague in detail, these reports proved enticing enough to convince Ed Purple to cancel other plans and organize a wagon train expedition to search for rich diggings in the high mountain country then known to only a handful of white Americans.

Beginning with his departure from Salt Lake City, Ed Purple's previously unpublished manuscript describes his 1862 expedition and his activities during the year that followed, a time spent in the rarified society of the Rocky Mountain West at the beginning of the Montana gold rush. The account is valuable not because the author was a person of great charisma, remarkable abilities, or stunning insights. In professional achievement, business acumen, intuitive shrewdness, or even flair for literary expression, Ed Purple lacked any truly distinctive talents. Yet precisely because he did not stand out among his contemporaries his narrative has a special claim on our attention. More representative than unique, Purple was part of a generation of young men who, lured to the West during the California gold rush, participated in opening other western regions to large-scale immigration and commercial enterprise. He provides a clear example of the continuity of pioneer experience in Montana and throughout western America during the latter half of the nineteenth century.

Like thousands of other hopeful gold seekers from many parts of the world, Edwin Ruthven Purple had joined the rush to California with every expectation of gaining a quick fortune. Born in upstate New York, the third son in a respectable family of English origins, Purple was seventeen years old and working as a clerk in a small dry goods importing firm in New York City when news of the gold discoveries in northern California reached the East Coast in the fall of 1848.[1] His employer soon decided to close the New York store and reestablish his business in San Francisco, the mercantile capital of the newly properous gold region. Before departing, he persuaded his youthful clerk to remain in New York to settle accounts and then join him in California. Accordingly, in mid-April of 1850 Purple boarded a steamship for Chagres, taking the speedy (and dangerous) Panama route across Central America. Three months later he arrived in San Francisco, where

---

1. The principal source of information regarding the Purple family and Ed Purple's career is a small volume entitled *In Memoriam*, prepared and privately printed by his brother, Dr. Samuel S. Purple, in New York City in 1881. Copies of this work are located in the New-York Historical Society and in the Montana Historical Society Library, among other repositories.

he confidently expected to enter business with his merchant friend amidst California's flush times.

These plans came to an end almost as soon as Purple set foot on dry land. Purple's one-time employer, having found the cost and risks of doing business too high in San Francisco, had already sold his stock of goods and sailed home with the profits. With scant resources, the young man found it necessary to shift for himself.

In these circumstances, Ed Purple's storekeeping experience pointed him toward profitable trading opportunities rather than mining. He soon made his way to Sacramento. There a friend helped him hire an ox-team and wagon, which Purple loaded with flour and other provisions and then took into the Sierra Nevada, following one of the overland emigrant trails to meet the wagon trains coming from the Midwest. His trading tour in the mountains, a romantic escapade as he subsequently recalled it, made him one among many petty entrepreneurs who in 1850 welcomed trail-worn travelers with emergency food supplies and alcoholic cheer at high prices. The enterprise, both financially successful and gratifying to young Ed Purple's spirit of adventure, served as a precedent for the later speculative project that would take him to the Montana goldfields.

Ed Purple remained in northern California for nearly a decade, easily adapting to the changeable opportunities of western life. Moving to the gold camps of Calaveras County, he supported himself as a miner and storekeeper, studied law briefly in the office of a San Andreas lawyer, and won election as a justice of the peace. For a year he served as a county supervisor and in 1855 gained admission to practice law in the local courts of Calaveras County. He also became a partner and secretary-treasurer in the San Antonio Ridge Ditch and Mining Company, one of the county's pioneer water development and lumbering operations. In these ways,

Purple supported himself without ever running the risk of becoming wealthy.

By 1860 California's gold mining economy had entered a period of transition, and the days of prosperity for small-scale placer operations vanished forever. In May of that year Purple became an agent with the Butterfield Overland Mail Company and moved to Fort Yuma. At that post, placed strategically on the southern route between Tucson and Los Angeles, he became the supervisor of the company's local operations.[2]

A year later the Lincoln administration negotiated a new contract with the Butterfield organization for overland mail service by way of a central route between Salt Lake City and Placerville, California. Purple, assigned to deliver a large contingent of the company's horses and stagecoaches to Salt Lake City, made the trip from Los Angeles—a journey of some eight hundred miles, much of it through desert country. With thirty men under his supervision, Purple traveled to Salt Lake in May and June of 1861. This adventure, completed without loss of a single animal, demonstrated the capabilities that later enabled Purple to lead his own expedition to the Montana country.

Purple remained in Salt Lake City as the chief agent for the Butterfield Overland Mail for nearly a year. The capital of Utah Territory and headquarters for Brigham Young's Church of Jesus Christ of Latter-day Saints, this city was also the principal business hub on the overland route between St. Louis and California. Although Purple remained unmarried, an observer might suppose that he had adopted a way of life

---

2. The federal census enumerator identified Purple in the summer of 1860 at Fort Yuma, in the Colorado Township of San Diego County, as one of ten single men living in a household headed by Louis J. F. Yeager, a thirty-five-year-old merchant from Pennsylvania. The other residents of the Yeager household included four Mexican Sonorans, one California Indian, and natives of Ireland and Germany, as well as a Virginian. 1860 Federal Census Rolls, Colorado Township, San Diego County, California, p. 812: NARS microfilm publication M-653, roll 64.

Salt Lake City, circa 1862 (Church of Jesus Christ of Latter-day Saints, Salt Lake City)

dangerously close to a settled, unadventurous, humdrum maturity in what was arguably the most morally respectable of all western urban centers. But events proved otherwise, for Purple was not yet ready to give up the quest for quick wealth.

Early in spring 1862, as Purple recounts, various reports reached Salt Lake City about rich gold discoveries in a remote area of the northern Rocky Mountains along the Salmon River, part of a region that would soon be organized as Idaho Territory. At the time, Purple had already made up his mind to leave the Butterfield firm, contemplating a long-deferred visit to his home in New York State. The accounts of the Salmon River mines, however, combined with information about the dangers of travel through war-torn Missouri, persuaded him to change his plans and try once again for riches in a new mining country.

In June he and a partner, C. L. Tisdale, purchased twenty-five oxen and three heavy wagons, then filled the wagons with trade items. Carrying flour and other staple foods, hand tools, and locally made cheap whiskey, they joined forces with a small collection of other adventurers and took the trail north from Salt Lake City, heading for a land that few white Americans had yet seen. After making a brief side trip to the former site of the Mormon Church's Fort Lemhi mission, Purple and his companions changed their destination from the Idaho mines to Deer Lodge valley, close to the Continental Divide in what later became western Montana. As Purple explains, they needed to halt before winter arrived, and they were impressed by the possibilities for finding gold in that region.

On the road to Deer Lodge, as chance would have it, Purple almost joined a prospecting party headed for one of the tributaries of the Beaverhead River. Because he and his fellow travelers did not know any of these prospectors, Purple and his companions continued on the Deer Lodge trail with their wagons. Doing so they again missed the big bonanza that was every gold hunter's dream, for the other prospecting party found rich placer deposits on Grasshopper Creek, where Bannack City soon arose, initiating the gold rush to western Montana during the late summer of 1862.

Purple, meanwhile, settled briefly in the Deer Lodge valley, established a store with the goods in his wagons, and began the tedious, muscle-straining labor of working the placer gravels along Gold Creek. But the gold deposits soon played out, while stories of richer discoveries from Grasshopper Creek and other new diggings lured away most of the camp's residents. After a brief, unrewarding trip to a prospective new gold camp in the Big Hole region, Purple returned to Gold Creek and remained there through a dull fall season. He then finally joined the migration to Bannack City, where he set up a store in December 1862 directly across the street from a saloon that was headquarters for the camp's roughest element.

With this move, Ed Purple became a firsthand observer of Bannack City's development during its boom era, a period of high expectations, reckless living, and intense social turmoil. For all of the 1863 mining season he remained at Bannack City, running his store and taking part in a few speculative ventures. In December 1863 ill health forced him to leave the gold camp and return to New York City, where he arrived in February of 1864. As he made this long-delayed journey home, he missed the start of the now-famous Montana vigilante movement that unfolded in Bannack City and Virginia City between mid-December of 1863 and early February of 1864. He learned later of the secret efforts of some of his Bannack friends and business associates, along with other determined townsmen, to destroy a gang of road agents and drunken rowdies headed, as they believed, by Sheriff Henry Plummer.

Although Purple originally intended to remain in the East only for a month or two, he never got back to Montana. In partnership with Governor James Doty of Utah Territory and others, he tried briefly to interest eastern investors in Montana mining property. Then, toward the end of April, he departed for the Rocky Mountain country, but severe illness overcame him. He returned to New York City in December with his health badly impaired. Purple still remained interested in Montana affairs, welcoming old friends from Bannack City whenever they appeared in New York, but his seasons of travel and exposed outdoor labor were behind him.

Nathaniel P. Langford's history of the Montana vigilante movement, published in 1890, gives Ed Purple a central role in a memorable incident during this period that forms an epilogue to Purple's Montana career. In summer 1869, Langford recounts, while he was visiting New York City, Purple requested that his friend Langford accompany him to call upon the brother and sister of Henry Plummer, the Bannack City sheriff hanged by the vigilantes as the alleged ringleader of the road agent gang. The brother and sister, Purple informed Langford, had been misled by Plummer's letters in 1863 to believe that he was executed because of his patriotic attachment to the Union cause. "They mourned his loss not only as a brother but as a martyr to the cause of his country."[3] To pursue and punish his murderers, as Purple and Langford learned when they made their call, the two grieving siblings had resolved to leave almost immediately on a trip to Montana.

According to Langford, he and Purple used all the arguments they could summon to convince the two Plummers to

---

3. Nathaniel P. Langford, *Vigilante Days and Ways: The Pioneers of the Rockies* (New York: A. C. McClurg & Co., 1912), 181–85.

abandon their plans, although—in an exercise of Victorian decency—they avoided revealing the true causes of their brother's death. When all else failed, the two men finally gave the brother and sister a copy of Thomas Dimsdale's vividly written book, *The Vigilantes of Montana*, which makes the strongest possible case for the justness of the vigilante movement and the iniquities of Plummer and his gang. In leaving this volume with Plummer's relatives, Purple and Langford assured them that it contained a completely true account of the events leading to their brother's death.

The following day the two men returned to the Plummer residence. They were greeted at the door by Plummer's brother, who reported that his sister was prostrate with grief at the revelation of Henry Plummer's criminal career. Although he also spoke in a voice broken by sobs and sighs, Langford concludes, the brother thanked him and Purple for allowing Plummer's family to discover the truth before they had pursued their misguided quest for vengeance all the way to Montana.

Although few details can be found, Purple apparently gave his attention to business in New York City during the next few years. Other concerns also came to occupy his time. In 1868 he married Mary Frances Close of New York. From this marriage came five daughters, three of whom survived into adulthood. About the same time Purple developed an abiding interest in genealogy. He joined the New York Genealogical and Biographical Society in 1869, the year of the society's formation, and his genealogical researches into the early Dutch families of New Netherland subsequently gained him local renown. He began an equally demanding task in 1876, when he became the acting librarian of the New York Academy of Medicine, undertaking to organize a large reference library donated by his brother, Dr. Samuel Smith Purple, one of New York's leading physicians and president of the Academy.

Ed Purple pursued this work until February of 1878, when his health broke entirely. He then resigned his post with the academy and returned to the directors a check for his entire salary during his term of service, funds that he had never spent. Ed Purple died in New York City in January 1879 at age forty-eight, a victim of pneumonia.

Amid his other bookish concerns, Purple at some point undertook to write a description of his Montana experiences, intending the account only for family reading. Like his Montana contemporary and friend, Granville Stuart, he had apparently kept a journal during his travels. (The narrative's specificity of detail regarding geography, chronology, and events is strong presumptive evidence for the existence of such a journal.) When he began to describe his adventures in Montana, starting with the trip northward from Salt Lake City, Purple adopted the journal format as the basis for his story, adding various remarks, explanations, and observations as he found them appropriate.

By the time of his death, Purple had completed most of his Montana manuscript, and he had gone over the draft at least once to make editorial amendments. He also placed with the narrative a brief encomium to the American miner, a set piece of Victorian prose that appears in this volume as an epilogue. Yet his account is incomplete; it breaks off in September of 1863, just after the robbery of Ben Peabody's stagecoach on the road from Bannack City to Virginia City. We do not know the reason for this abrupt stop, but it may well have been due to Ed Purple's failing health.

Purple's incomplete Montana manuscript, along with other personal papers, at some point came to rest in the New-York Historical Society in New York City. It is logical to assume that these materials were donated by his widow or his brother as literary executor, but the society has no record of the gift. The original journal, if it survived, has not come to light. The New-York Historical Society has cooperated fully in the arrangements for editing the manuscript and its publication by the Montana Historical Society Press.

Close textual examination and editorial efforts to confirm the accuracy of Ed Purple's narrative reveal certain characteristics that readers should consider. First is the matter of dates. Although Purple attempted to give his trail narrative, in particular, the appearance of a consecutively dated journal, in fact he was not completely accurate. Cross-checking, as indicated by editorial notes accompanying the narrative, shows a variation of four to ten days—and sometimes more—between the dates he set down and the true dates for events that can be confirmed by other primary sources. Presumably, the journal Purple used as his source was itself not dated, and during the transcription process our author merely made approximations of what seemed reasonable dates.

On geographical matters the narrative is accurate. Where they can be checked, Purple's estimates of distance are roughly correct. In retracing his route and visiting sites that Purple saw in 1862 and 1863—a task undertaken by the editor during the summer of 1993—it becomes apparent that the author was a careful observer and recorder of the landscape, although his memory and romantic imagination may have enhanced in retrospect a few scenes that he describes. (Results of the editor's travels and on-site visits are recorded in notes accompanying the narrative.)

The identification of individuals mentioned in the narrative became a matter of substantial concern in preparing this work for publication. Simply put, Ed Purple had almost an obsessive preoccupation with names. He demonstrates in his manuscript account an enthusiasm for recording the identities of people he met that helps to explain his subsequent devotion to genealogical research. With the assistance of many colleagues, friends, and new acquaintances who became collaborators in this project, the editor has been able to verify the existence of most of these individuals and provide for them a broader biographical context. In every case save two or three, Purple's information about these people has proven correct. Only a few persons important to Purple—unfortunately including two who were closely associated with him in California and later in Montana—are undocumented in other sources examined by the editor, and thus remain obscure in these pages.

With its essential accuracy confirmed, Ed Purple's narrative can be considered historically valuable from two perspectives, one local and specific, the other more general in character. In local terms, Purple gives us a broad-spectrum report about events and episodes in the northern Rockies during an era that is distant in time but still immediate in interest. In relating the particulars of his wagon trip northward, he supplies details not found in other accounts from this period. Among other items of information, he records a nearly disastrous attempt by a crowd of would-be miners to find riches near the Mormon church's abandoned Fort Lemhi mission in east central Idaho. His narrative contributes to our knowledge of the events surrounding the discovery and initial development of western Montana's gold resources and enhances our understanding of social and cultural relations, including Indian-white relations, during the earliest phase of the Rocky Mountain gold rush.

For Montana history enthusiasts, Ed Purple makes a particularly important contribution with his detailed account of life in Bannack City and environs during the first year of the town's existence. His narrative places particular emphasis on the problems of law and order at Bannack City, an emphasis undoubtedly inspired by his knowledge of the vigilante movement that took shape a few weeks after he departed Bannack City. Unlike Thomas Dimsdale, whose secondhand account has been widely accepted as the prime authority for this historical era, Purple observed at firsthand the preliminaries to the vigilante outbreak. His store was directly across the street from Skinner's saloon, the principal gathering place for the worst of the socially disreputable and dangerous characters who frequented the town. Moreover, as we learn from another documentary source, Purple served as an election judge in the voting that made Henry Plummer the sheriff of Bannack City.

Clearly, Purple was an independent observer, not tied closely to any of the social cliques that could be found in Bannack City in 1863. Like Plummer and his close associates, Purple was a veteran of the California gold rush, but he did not regard these men as compatriots. Like many of the men who would league together as vigilantes, he was a member of the Masonic order, but he did not select his associates on the basis solely of this fraternal bond. A New Yorker by birth, he could not be identified with the Minnesota "Yankees," who formed one large element in Bannack society, nor did he find a place among the predominantly border state and southern "Pike's Peakers" who arrived in Bannack City fresh from the Colorado gold rush. In short, Purple was a mining-camp merchant, with his store open to all and his friendships apparently based on ecumenical principles.

As Langford's story from 1869 demonstrates, Purple was quite familiar with Thomas Dimsdale's book, published in 1865, at the time he wrote his own narrative. Because his partial retelling of the story comes from an independent, firsthand observer, historians will conclude that Purple's account adds substantial weight to the case against Henry Plummer and his alleged partners in crime. Purple confirms that the relative lack of community sentiment and the related tolerance for criminal behavior had become a major problem in Bannack City by the fall of 1863. He adds detail, and his narration differs with both Dimsdale and the later-published Langford account on some relatively minor matters. But, in his descripton of events and his portrayal of Henry Plummer's character, Ed Purple emphatically substantiates the substance of the case against Plummer and his associates first made in print by Thomas Dimsdale. While we can only speculate about what further evidence Purple might have contributed to the historical record had he lived to finish this narrative, his incomplete work makes still more tenuous the recent attempt by authors R. E. Mather and F. E. Boswell to strike, as it were, a historical plea bargain on behalf of Plummer and other men hanged by the Montana vigilantes.[4]

Although there is no reason to believe that he was conscious of it, Purple wrote within a well-defined tradition, the true tale of frontier travel and adventure. Considered in a literary context, Ed Purple's account can be seen as an excellent example of the gold-rush narrative, reflecting a classic episode in the history of western America. With its emphasis on unusual sights, unusual individuals, and unusual incidents during his journey northward and the time he spent in Montana, Purple's literary effort, like others in this tradition, implicitly glamorizes the life that the author had once known in the West. But, unfortunately for some current historical interests, the tradition did not give attention to the gritty, grimy details of day-to-day labor: Purple tells us virtually nothing about his activities as a miner or about the business of running a store in Gold Creek and Bannack City.

Anyone versed in recent historical scholarship concerning the American West can discover in the pages of Ed Purple's Montana gold-rush narrative many other general themes that are pertinent to contemporary concerns. Directly and indi-

---

4. R. E. Mather and F. E. Boswell, *Hanging the Sheriff: A Biography of Henry Plummer* (Salt Lake City: University of Utah Press, 1987).

rectly Purple addressed issues of Indian-white relations, the meeting and mixing of different European American ethnic and cultural traditions, the prevalence of public violence, and the shaping of social consensus within unsettled new communities—all matters that have become pertinent to our ongoing exploration of the frontier elements in our national heritage. He even provides a few snippets of information and commentary concerning the circumstances of women in the social milieu of Deer Lodge in 1862 and Bannack City in 1863.

While Purple does not write in detail about business, we should not doubt his concern for making money. From the time he and his partner C. L. Tisdale decided to invest their savings and head north from Salt Lake City in a highly speculative enterprise, however, his actions tended to stem more from adventurous ambition than prudent calculation. Today we might suppose only a type of midlife crisis would lead such men to risk their property and their lives in a last, blindly hopeful effort to strike it rich in the supposed new goldfields of the northern Rockies.

Yet in the end it is perhaps an image of Ed Purple the author, not the adventurer, that should be most enduring. Purple's prose acquaints us with a complex individual, capable of dreamy soliloquies about the natural beauty of the Rocky Mountain landscape and musing about the vagaries of human nature. Back home in New York at last, comfortably surrounded by family and friends, while his health prevented him from ever again traveling west, surely he gained satisfaction in setting down for posterity this account of more stirring times in far-off places.

EDITORIAL NOTE

In preparing Ed Purple's narrative for publication, I have not attempted to reproduce exactly the appearance of the original. As he wrote and rewrote some passages, Purple crossed out sections, interlined, inserted words, and changed phrases. Adding and deleting as any careful writer will do, he left a rather messy manuscript. His habits of capitalization and punctuation were somewhat haphazard, not conforming to the customary practices either of his time or ours. Moreover, though his writing shows the marks of a sound education and wide reading, Purple displayed a few persistent idiosyncracies in spelling. The word "wagon," for example, he usually spelled "waggon," the word "Mormon" he always represented as "morman," and he failed to capitalize the word "Indian."

As an editor, I have adopted a moderate course of textual emendation. For the most part I have retained Purple's own capitalization and his use of ampersands, allowing these matters to stand uncorrected in order to provide a certain sense of the original. Errant punctuation I have generally altered in accord with current standards, making the text more consistent and clearer for modern readers. At a few other points, simply for clarity's sake I have silently rectified Purple's spelling errors. Solely to reflect my editorial preference for avoiding personal offense, I have also used the corrected forms for "Mormon" and "Indian." Brackets set off all insertions. Otherwise the text appears substantially as Ed Purple left it, incorporating the revisions that he had indicated in the manuscript.

CHAPTER ONE

✦

Salt Lake City to the Snake River Ferry
June 7 to July 5, 1862

Early in the spring of 1862 reports reached Salt Lake City through the California Papers and other sources, of the Discovery of rich deposits of Gold on the Salmon River in Oregon. The most moderate accounts represented these Gold Placers as of unparalleled richness but of limited extent.[1] I had determined to quit the employ of the Overland Mail Company with the intention of going home by the commencement of summer. But the war of the rebellion then [being] fairly inaugurated, and the consequent disturbed condition of affairs along the line of travel after I should have reached the Missouri River, led me to give up the idea. I then concluded to go to the Salmon River Mines.

---

1. The Salmon River mines formed an isolated district of placer diggings centered on the towns of Millersburg and Florence in the Salmon River basin, north of the Salmon River in modern Idaho. Discovered late in the 1861 season, these diggings attracted gold seekers from every direction in 1862, but the deposits quickly played out. Only at Florence and Warren's Diggings did significant mining continue through the 1864 season. For the story of this gold discovery and the first rush into the area, see Hubert Howe Bancroft, *History of Washington, Idaho, and Montana, 1845–1889*, in *The Works of Hubert Howe Bancroft* (39 vols., San Francisco: History Company, 1890), 31:244-59; and William S. Greever, *The Bonanza West: The Story of the Western Mining Rushes, 1848–1900* (Norman: University of Oklahoma Press, 1962), 259-61. Officially a part of Washington Territory—not Oregon—since 1853, this area remained innocent of formally authorized government until after Congress created Idaho Territory in March 1863.

For this purpose in company with C. L. Tisdale I purchased an outfit consisting of 25 Head of work oxen & 3 Government waggons.[2] We loaded my waggons principally with Flour & Bacon, with a good supply of Coffee, Tea, Sugar, Salt, Dessicated Potatoes and other vegatables, Picks, Shovels, and such tools as we could pick up in Salt Lake City, which we thought we should want in our journey, and in a brand new settlement. A number of the old employees of the Overland Mail Company, some of whom had accompanied me from Los Angeles the year before, hearing of my intended expedition to the mines, were desirous of going with me.

On the 7th day of June, 1862, our train, Consisting of 6 waggons and 13 Persons, left Salt Lake for the Salmon River country.[3] I remained in Salt Lake City two days after the train had left for the purpose of settling up some business connected with its outfitting, and on the third day after its

---

2. In the parlance of the time, government wagons were army-designed freight wagons, heavily built and usually pulled by teams of six oxen or mules. After the Buchanan administration withdrew federal troops from Utah Territory in 1858, at the end of the so-called Mormon War or Utah War, the freighting firm of Russell, Majors, and Waddell auctioned off a large number of these vehicles at Salt Lake City for ten dollars or less a wagon. See Nick Eggenhofer, *Wagons, Mules and Men: How the Frontier Moved West* (New York: Hastings House, 1961), 113-28; and Alexander Toponce, *Reminiscences of Alexander Toponce, Written by Himself* (1923; new ed., Norman: University of Oklahoma Press, 1971), 24-25.

3. The following list of the members of this party appears on the reverse of the first page of Purple's manuscript: "C. L. Tisdale, S. W. Batchelder, William Cole, John Murphy, Dr. Ford, Horace Wheat, Joseph Vassar, P. C. Woods, John Rhinehart, George Dewes, James Gamble & Son, Henry Gilbert & Brother in Law." Not counting Purple but including the two unnamed individuals, the total is fifteen rather than the thirteen specified in the narrative. Of these men, Tisdale, Batchelder, Cole, Wheat, Woods, Rhinehart, and Dewes all later appear as residents of Bannack City in a census of non-Indian settlers in Montana during the winter of 1862-1863 prepared by Purple and Wilbur Fisk Sanders, and published in *Contributions to the Historical Society of Montana* (hereafter *Contributions*) (10 vols., Helena: Rocky Mountain Publishing Co., 1876–1940), 1:334-48. Both this listing of his party members and the Bannack City census demonstrate Purple's keen concern for recording the names of persons with whom he associated, a concern evident throughout his manuscript.

Beside the flagpole in this photograph, taken about 1863, stands Salt Lake City's Overland Mail Company, the firm that acquired the business and property of the Butterfield Overland Mail soon after Ed Purple's departure. (Church of Jesus Christ of Latter-day Saints, Salt Lake City)

departure, joined it in the evening at the Bench creek 30 miles north of the city. This was the 10th of June.

Our route was the northerly road leading out of Great Salt Lake city, passing the warm spring just out of the walls of the city, in which I had so often bathed, and which had relieved me effectually of the Rheumatism, from which I had suffered much at Fort Yuma and Los Angeles.[4] On the right of us rose the peaks of the Wasatch range, sparsely spotted with dwarfed and gnarled mountain cedar. On our left stretching northward, lay the Great Salt Lake with its deep sky blue water, its shore lined with white incrustations of salt, resembling a huge emerald in a silver setting, as its waters danced in the sunlight of that bright June morning.[5]

On the 10th of June, 1862, Robt. T. Burton, the sheriff of Salt Lake City, was with a posse of men at Weber Creek, sent by the authorities of the Mormon church to arrest [Joseph] Morris, Bangs [John Banks] and others of the sect of Morrisites, who had split off from [the Mormon Church], and set up an independent Church Government. These Morrisites were communists, dividing and sharing with each other their wheat, Flour, cattle, and property of every descrip-

---

4. A favorite place of resort during the early decades of Mormon and Gentile settlement in the Salt Lake valley, the Warm Springs flowed from the foot of Ensign Peak one and a half miles from Temple Square on the main route leading northward from Salt Lake City. Brigham Young and a few of his colleagues first enjoyed bathing in these waters two days after the Latter-day Saint pioneer party entered the valley. A public bathhouse, authorized by the territorial legislature in 1848, became the center of an entertainment and social complex that, as a Salt Lake City newspaper phrased it in 1866, would draw crowds from the city "in the still summer evenings, for health, recreation, and innocent pleasure": *Deseret News*, June 28, 1867. With later improvements in facilities, the Warm Springs remained open into the 1930s. Among other accounts, see particularly Andrew Jenson, ed., *The Historical Record: A Monthly Periodical, Volumes V, VI, VII, VIII* (Salt Lake City: Andrew Jenson, 1889), 339-40; Kate B. Carter, comp., *Our Pioneer Heritage* (20 vols., Salt Lake City: Daughters of Utah Pioneers, 1958–1977), 8:439-40; and John Hudson, *A Forty-niner in Utah, with the Stansbury Exploration of Great Salt Lake: Letters and Journal of John Hudson, 1848–50*, ed. Brigham D. Madsen (Salt Lake City: Tanner Trust Fund, University of Utah Library, 1981), 84, n. 25.

Salt Lake City sheriff Robert T. Burton, shown here circa 1890, accompanied the posse that killed Morrisite leader Joseph Morris in the sect's confrontation with Mormons at Weber Creek. (Church of Jesus Christ of Latter-day Saints, Salt Lake City)

tion. They differed from the Mormons in Polygamatic habits, having only one wife. Joseph Morris, a man upwards of fifty years of age, and an early member of the Mormon Church, was their Prophet & Leader.

So great was their religious infatuation they entirely neglected the sowing of crops this spring, Morris having prophicied that in that month Jesus Christ would make his appearance on a white Horse and, coming down from the top of a certain hill near by, where he would first be seen, into the Weber valley where they lived, would lead them down to Salt

---

5. Heading northward from Salt Lake City toward Fort Hall on the Snake River, the Purple and Tisdale wagon train had a well-defined track to follow; immigrants, freighters, and other travelers had been using the route since early Mormon settlement in the Salt Lake basin in 1847. In general the route paralleled the line of modern Interstate 15 to the vicinity of today's Brigham City, where it turned northwest to reach a ferry crossing over the Bear River, then proceeded up the Malad River valley to the locality of modern Malad City. At that point the customary trail continued northward along the valley of the Little Malad River, crossed the Bannock Range and then the lower Portneuf River near Pocatello, and proceeded through open country to Fort Hall. Because Purple and Tisdale were making the trip early in the season, they followed their guide's advice and detoured from the established route before reaching the Little Malad River to avoid high water on the lower Portneuf River.

Lake City and give into their hands the property and posses-
sions of the inhabitants; and place Morris at the head of the
Church, in place of Brigham Young. If Brigham resisted (as
fearing neither God, Man, or the devil, he might do), the city
would be vented with the besom of destruction and those
who opposed Christ their leader, would be put to death by
the sword. The order of this procession was minutely laid
down, Jesus Christ going first, Joseph Morris following
immediately after, on horseback with a drawn sword, and the
Morrisites some Two Hundred in Number, following in the
order of the official rank they held in the Church.

 This prophecy was soberly told me in Salt Lake City, by
Bangs who was next in authority to Morris, in the month of
May 1862. About the ordinary affairs of life, unconnected
with Religious Matters, Bangs was a sensible old man, but I
need hardly add he was a confirmed Monomaniac as well as
all the balance of the Morrisites, who ought [to have been]
and had they lived in any other Country would have been
shut up in a Lunatic Asylum.

 Morris & Bangs had several interviews with Brigham in
the spring, in one of which it was reported that Brigham told
Morris he would kick Jesus Christ's behind out of his Lion
house, if he came there with him, Joseph Morris.

 It was said that the Morrisites, who had become by their
habits very destitute and poor, had stolen from their
neighbours, on the Weber, Wheat, Flour, stock, and other
property, and this was the ostensible purpose of Burton's and
his posse's visit to them. I did not go down to the settlement,
although [it was] in site of the road I was travelling. I suppose
I was the only *Gentile* near the scene of disturbance and death
that day.

 The result was, Morris was killed, Bangs taken prisoner
and the sect completely dispersed by that day's proceedings.[6]
Afterwards they moved to Soda Springs on Bear River and

were protected by Genl. Connor's troops after their arrival in the Territory.[7]

On the 11th [of June] we camped at Ogden Hole, having in that day crossed the Ogden River just above the town. The River was much swollen by recent heavy rains, and our waggons were Ferried on a small Ferry skiff, owned by a Mormon. We had to unload all our waggons and Ferry our provisions in lots of 1800 to 2000 lbs. each separately. Our cattle & horses were made to swim, and with much difficulty we got them safely across, without losing a single head.

On the 12th we camped near the warm springs four miles from the Hole. These springs are unlike those near the city of Salt Lake, being more strongly impregnated with Iron, with less salt and sulphur. Their temperature at the head is near the boiling point.

On the 13th we camped near Empy's house.[8] It had rained nearly every day since our departure from Salt Lake City, and on the 14th & 15th the rain was so incessant we were not able to travel, and remained at Empy's. We found

6. The fatal confrontation at Kingston Fort actually took place on June 16. Most likely Purple passed the locality before these critical events took place, although it is possible that he was writing entirely from memory without the benefit of a trail diary or other record, and that his recall of dates was erroneous. On the Kingston Fort conflict and a detailed, balanced history of the Morrisite movement generally, see C. LeRoy Anderson, *Joseph Morris and the Saga of the Morrisites* (Logan: Utah State University Press, 1988), esp. 101-45.

7. Colonel Patrick E. Connor took command of the U. S. Army's Military District of Utah, Department of the Pacific, in August 1862, bringing from northern California the volunteer Third Infantry Division and part of the Second California Cavalry. Promoted to Brigadier General in March 1863, Connor remained in command at Salt Lake City until he left the service in April 1866. His Utah career, marked by antagonism to Brigham Young and the Latter-day Saint leadership and by a hard-hitting campaign against the northern Shoshones, is described most fully in E. B. Long, *The Saints and the Union: Utah Territory during the Civil War* (Urbana: University of Illinois Press, 1981); and Brigham D. Madsen, *The Shoshoni Frontier and the Bear River Massacre* (Salt Lake City: University of Utah Press, 1985), 143-46, 164-69, 177-200. For Connor's protection of refugee Morrisites at Soda Springs, see Long, *Saints and the Union*, 180-81, and Anderson, *Joseph Morris*, 165-71.

here a young man who had been badly wounded by the Indians. He was coming east from California with a small party of six men including his brother. Lortening [loitering] behind his party, a day's travel from Bear River, he had met three Indians who wanted to trade some buckskins for caps and powder. He foolishly bartered with them, and when [he was] turning to leave, they discharged their arrows at him, one of them striking him above the small of the back, inflicting a dangerous wound. When inquiring a year and a half later, I learned that he recovered, although he laid three weeks in bed before being able to get about.

On the 16th we reached Bear River. We were overtaken today by a man named John Jacobs who with his Indian daughter, a girl of ten years, and [a] horse packed with flour, was on his way to the Deer Lodge Country, where he lived.[9] As he was familiar with the route and country we were to travel after leaving Bear River, I offered to put his flour & little girl in our waggons, in lieu of his services as a guide. He

---

8.   Most likely this person is William Empey, who arrived in Utah as a member of Brigham Young's pioneer party in July 1847. In 1852 the territorial legislature granted to Empey and three partners a franchise to operate a ferry across the Bear River. According to a newspaper account, he was still running the Bear River ferry in December 1862, although he may have sold the business the following year. See Daughters of Utah Pioneers of Box Elder County, *History of Box Elder County* (Brigham City: Daughters of Utah Pioneers of Box Elder County, 1930?), 147-48; (Salt Lake City) *Deseret News*, December 17, 1862, quoted in Madsen, *The Shoshoni Frontier*, 174.

9.   A longtime resident of northern Rockies, John M. Jacobs joined Jacob Meek, Robert Dempsey, and Robert Hereford in trading cattle and horses with emigrant parties at Fort Hall in 1856. With his stock, he traveled north to the Beaverhead valley and settled there with his native wife and family. He receives frequent mention in Granville Stuart's *Forty Years on the Frontier as seen in the Journals and Reminiscences of Granville Stuart*, ed. Paul C. Phillips (2 vols., Glendale: Arthur H. Clark Company, 1957). After guiding Purple and Tisdale's train north into the Montana region (and still accompanied by his young daughter), Jacobs joined John Bozeman in exploring and opening the wagon route from the North Platte River that was called first the Jacobs and Bozeman Cutoff, and later the Bozeman Trail. For a biographical sketch, see Dan L. Thrapp, *Encyclopedia of Frontier Biography* (3 vols., Glendale: Arthur H. Clark Co., 1988), 2: 718-19.

had lived many years with the half breeds in Deer Lodge and vicinity, with a Flathead Indian woman, the mother of the little girl, Emma. He was a coarse fellow, and brutal to his child, cuffing, kicking, and striking her with anything that came to his hand, for the least disobedience. We had to frequently check him in these outbursts of passion, fearing that he would commit some serious bodily injury to her.

On arriving at Bear River we found incamped there three Mormons, Christopher Stokes, John Tripp and his Son, John Tripp, Jun.[10] They had their waggons loaded like our own with provisions, and like ourselves were on their way to the Salmon River Country. The heavy rains which had fallen since we left Salt Lake City had made the roads on the low ground almost impassible for our heavy Government waggons, and we concluded to encamp for a few days, as every pleasant day improved the condition of the roads wonderfully.

We made our camp in the Sage brush, on the bluff a few yards from the river, near the California and Oregon [trail] crossing. There was no trees, and our meals of bread and bacon were eaten under the shade of our waggons. The mosquitoes annoyed us terribly, and the only relief we could get from their attacks was to build fires or smudges of dry manure about the camp, and keep ourselves enveloped in the smoke as much as possible. My Poney would leave off grazing and come to the camp, and remain for hours in the smoke to free himself from these tormentors. The bite of the Mosquitoes was so poisonous to some of the men, that they were nearly blinded; in the morning theirs eyes and faces would be swollen fearfully. I was probably as much bitten as anyone,

---

10. Stokes and the senior Tripp later returned to Salt Lake City, where they both lived to old age. An obituary for Christopher Stokes appears in the *Deseret News*, January 26, 1916, p. 2. John A. Tripp died two years later. His obituary, *Deseret News*, February 13, 1918, p. 5, noted that he was survived by two sons.

but escaped from the serious effects it had upon the others. We were in danger too, in this camp, constantly, of being bitten by Rattlesnakes. These reptiles were more abundant here than any place I was ever at. It was not an unusual circumstance to kill six or seven of them, in and about the camp in a single day.

On the 19th of June I had a narrow escape from being bitten by one. Sam Batchelder and myself had gone to the river for the purpose of bathing. We placed a small piece of board near the water's edge to stand on when dressing. The river was swollen beyond its ordinary banks, and the water reached the sage brush. After my bath I got onto the plank to dress. I had just got my shirt over by head when I heard the sharp rattle of a snake, and instantly sprang into the water. Upon reconoirting [reconnoitering] the position of the enemy, I found him coild up in a sage brush, within a foot of the place where I was standing on the plank, a moment before, stark naked. As he seemed determined to contest the right to the position in which he had ensconced himself, and as I could not prudently finish my toilet while he remained there, I got a stick and with Sam's assistance killed him after he had shown us considerable fight. He had seven rattles which I cut off as trophyes of the battle.

These snakes are by nature Magnanimous and courageous. They exhibit these qualities in the mode and method of their attacks upon the disturbers of their peace and quiet. They never without they are trodden upon, or in some way molested, offer to attack man or beast, and then only after having given notice with their rattle, if they have unconfined use of their tails. The noise they make with the horny joints at the end of their tails, called rattles, very much resembles the buzzing noise of a large grasshopper, when it raises itself from the ground for flight, and I have been startled a hundred times by these insects, by the similarity of sound.

On the 22nd of June 1862 we had our waggons ferried over Bear River. Our Cattle & Horses were made to swim. The River was, on account of high water, at least one hundred & fifty yards in width, and we had much difficulty in getting our stock all over safely. This we however succeeded in doing, and moved up the west bank, four miles above the Ferry, to Camp. Our party now consisted of the following named persons: C. L. Tisdale, Wm. Cole, Dr. Ford, S. W. Batchelder, John Murphy, Horace Wheat, John Jacobs & daughter Everina, Joseph Vassar, Henry Gilbert & his brother in law, and myself with five waggons. P. C. Woods, James Gamble & Son with one waggon. John Rhinehart, George Dewes and Harry Gray with one waggon. Christopher Stokes, John Tripp Sr. & John Tripp Jr. and John Knowles, with 3 waggons.[11]

As we were now fairly in the Bannack Indian Country, and liable to have our stock driven off by them, our Company elected me as Captain, promising implicitely to obey such orders as I might deem necessary for the safety of ourselfs and stock. I divided the Company into four guards of four each, to take charge of the stock each night when we encamped. The first guard went on duty as soon as our cattle were unyoked at night, and were relieved by the second guard at 12 o'clock M. The second guard brought up the stock in the morning by early daylight, and we were prepared for the day's travel. The 3rd guard went on the next night, relieved by the 4th at the same time as before. On the 3rd night the 2nd guard would take the first watch & the 1st the last. Alternating in this manner the fatigue of guard duty was made as light

---

11. Like his fellow-Mormons Stokes and the two Tripps, John W. Knowles was also apparently taking a wagon load of goods northward toward the new mining discoveries on speculation. He later returned to Utah and took up residence in Salt Lake City, where he died in December of 1920. His obituary appears in the *Deseret News*, December 10, 1920, p. 3.

as possible to each man. My Partner, Tisdale, had a large Newfoundland Dog, and Gamble's son a small physe [feist] that were faithful watchers of our own camp, and shared with two volunteers from the men, the duty of guarding the Camp when we slept.

Our Camps were generally selected with reference to their proximity to wood and water, and good feed for our animals. When forming a Camp at night our waggons were driven close together in circular form thus ⋂ something like a horse shoe. This answered the purpose of a Corrall, or pen in which to yoke up our cattle, and at the same time afforded us a protection in the event of being attacked by the Indians. Our party was well supplied with arms and ammunition, and able to have made a stout resistance if molested by them. The guns & side arms were kept in good order, as most of our party had been enough in Indian Countries to know that it would never do, to go loitering carelessly through one, with which they were unacquainted, as we were with this.[12]

A report had reached us before we left Salt Lake City that Jake Meeks, who had come down from the Deer Lodge Country to establish a Ferry near Fort Hall on the Snake river, had been attacked by a party of Bannack Indians, just ahead of where we then were, and that his party had killed two of

---

12. These precautions were well advised. Shoshone hostilities disrupted the overland mail and stagecoach service west of Salt Lake City in the spring of 1862, while Bannock and Shoshone raiders assaulted emigrant trains in a wide area between Soda Springs and Raft River by the time of Purple's trip northward. See Madsen, *The Shoshoni Frontier*, 133-55.

13. Formerly a mail and express carrier for the Hudson's Bay Company between Fort Hall and Fort Boise, Jacob Meek took up residence in the Beaverhead valley in 1856 and began trading with emigrant parties at Fort Hall on the overland trail. He guided James and Granville Stuart to the Beaverhead country in 1857. In a journal entry for April 10, 1862, James Stuart noted the planned departure of Meek and several other traders from the Deer Lodge country for the overland trail. Stuart, *Forty Years*, 1:204.

the Indians.[13] Jacobs spoke with some confidence of the truth
of this report, and this alone was sufficient to make us prudent and cautious in our manner of travel.

Just above the camp we made here, could be seen the
towering walls of Bear River Can[y]on, through which the
stream passes on its way to the Lake. On our right looking
northward lay the high ridge, at the base of which we had
travelled from Salt Lake City, and which at this point, less
rugged and precipatous in its proportions, dwindles down to
a succession of small hills that environ Cache valley. On our
left the country extended itself out in a level plane, and
bound the line of our vision when it reached the high ground
on the horizon, which Jacobs told me was the foot hills of the
Goose Creek Mountains. The Malad, a sluggish alkalie stream
of water, runs through this plain; it has washed its way
through the deep soil of the valley until the banks which
confine it are on a level with the surrounding country. From
this circumstance its waters cannot be seen at any great
distance. It flows as do all the other streams in the Great
Basin, into the Great Salt Lake.

On the 23rd we got a good early start, and rolled out on
our journey. The day was warm, but not unpleasant. We
killed two Curlews, which Sam Batchelder and myself broiled
at our camp fire at night for supper. They are not a shy bird, I
should judge. They make their nests in the grass like the
Meadow Lark. They have a Sharp schrill note when on the
wing, and their scream is more frequent just before rain. They
are about the size of a two months old chicken, with a long
bill, somewhat like a Woodcock, of a brown mottled plumage
and long legs. The meat was very good.

We were overtaken before noon by a Mormon, his wife,
and boy. I believe they are Morrisites.[14] They have two or
three cows and we had plenty of new milk for our Coffee and
some to drink.

The distance we travelled to day is estimated at 17 miles, and our camp for the night is just beyond the Utah & Oregon line, which is marked by a small stone slab "U & O Line".[15]

The 24th we were overtaken by Frank Troucheu [Truchot], a Frenchman, and his Indian wife, who lived at Deer Lodge.[16] He had pushed along rapidly from Salt Lake with his cattle and waggon, to overtake Jacobs, and seemed pleased at finding so large a company to travel with. His face was much swollen, and his eyes nearly closed, from the effects of Mosquitoe bites, on Bear River.

We camped tonight near the crossing of Deep Creek, a tributary of the Malad, having made about 15 miles.[17]

25th of June, we crossed Deep Creek. The crossing was difficult owing to the light soil of the bank, in raising up out of which, on the opposite side, our waggons would sink up to the axle trees. We cut ash and willow bows [boughs] and laid them down, which aided to support our heavy waggons somewhat. But it was hard work for our poor oxen, as well as ourselves.

---

14. Following the fight at Kingston Fort, a few Morrisite families left Utah Territory and headed north into the Deer Lodge valley. Encouraged by the self-designated new leader of the movement, George Williams, other survivors and fresh converts joined them in time, forming a loosely knit community that remained strong until the 1890s and endured into the 1940s. See C. LeRoy Anderson, "The Scattered Morrisites," *Montana The Magazine of Western History*, 26 (October 1976), 52-69; and Anderson, *Joseph Morris*, 206-29.

15. In reality the boundary then separated Utah and Washington territories, although officials at Olympia had done little to assert civil jurisdiction east of the Cascades.

16. A native of France, born in 1833, Frank Truchot had come to the United States in 1852. Working as a herdsman, he arrived in Utah in 1857 as a civilian employee for the army. Two years later he moved to the Deer Lodge country and made his living as a trapper, hunter, and trader. He also became the first in the area to attempt farming. In 1868, after a brief trip to California, he brought a herd of cattle from Utah and resumed farming and ranching near Deer Lodge. Subsequently he moved his operations to the Teton River country west of Fort Benton, where he still lived with a large family in 1885: Michael A. Leeson, *History of Montana, 1739–1885* (Chicago: Warner, Beers & Co., 1885), 1028.

17. Interstate 15 today crosses Deep Creek about three miles north of Malad City, Idaho.

*Frank Truchot of Deer
Lodge, circa 1880*
(Montana Historical
Society Photograph
Archives)

We made about 6 miles to day and camped at what we called Spring Valley Creek.

During the night we were visited by a furious storm of rain and wind, and most terrific thunder and lightning. It seemed as if we were about to be involved in a general wreck of Nature, by the fury of the elements. So Severe and terrible was it, that our Cattle tired and weary as they must have been from their hard day's work, acted as if they had gone mad, and in despite of all the guard could do, broke into a general stampede and fled over the country before the wind and rain. Some of them must have travelled at least 10 miles before the storm abated, and they could be overtaken.

We got them all together by 10 o'clock on the 26th and moved on to the head of a fork of Marsh or Swamp Creek, which flows into the Port Neuf, a tributary of Snake River. We had crossed to day the dividing ridge between the Salt Lake waters, and those of the Columbia River, and were now out of the Great Salt Lake basin. We had travelled about 6 miles this day.

In reaching this point, we had diverged at Deep Creek from the usual line of travel to Fort Hall over Bannack Mountain, which led farther west over "Devil & Satan" Creek, the

inelegant but suggestive name given it by the old Mountaineers owing to the difficulty of "getting around it" or "over it"
by reason of the quick-sands in which its banks and bed
abounded. Jacobs said the "devil & Satan" would mire a
Mackanaw blanket, and after you are over, you will have to
Ferry across the Port Neuf, and advised us to avoid it by
turning off towards Marsh & Swamp creek, and crossing the
Port Neuf River high up, where its rivulet size would offer
little or no obstacles to our transit.[18]

On the morning of the 27th, shortly after we had started
on our journey, I was surprised upon looking back on the
road we had travelled to see comming down a nearby hill
some fifteen or more waggons drawn by mules and horses,
accompanied by thirty or forty men, who were hurrying along
rapidly on our track. I had posted a notice at Deep Creek,
stating for the benefit of any partys that might be coming
along behind us, the reason of my turning off at that point, as
I thought the information Jacobs had given me would be
useful to others. I had however no idea of there being from
any quarter such a large emigration on the road for the
Salmon River Country as this unexpected sight seemed to
indicate.

Upon their coming up, they asked us where we had
outfitted from. We answered from Salt Lake City. They told
us that they were from Colorado, and instead of turning off at

18. After leaving the Malad River, rather than taking the customary route
northwest along the Little Malad River, Jacobs adopted a route north and
northeast that generally parallels the line of U.S. Highway 15 today north of
Malad City, moving over the Bannock Range by way of Malad Summit. To avoid
the lower Portneuf River crossing, he then took Ed Purple's party on a circuitous
detour farther to the northeast, cutting across Marsh valley and the Portneuf
Range to reach the Hudspeth Cutoff of the Oregon-California Trail route in the
vicinity of modern Lava Hot Springs, then headed east along the Hudspeth
Cutoff to intercept the main Oregon-California Trail a few miles west of Soda
Springs. Although Purple later came to doubt the wisdom of his guide's
directions, Jacobs demonstrated here a remarkable knowledge of the region's
geography and the network of alternative east-west trail routes.

Fort Bridger for Snake River, which would have been their most direct course, they had come round by the way of Salt Lake City.

As they were travelling with horses and mules, they traveled faster than we could with oxen, but the company kept near us until we reached Subletts Cutoff, when they pushed on ahead to Snake River.[19] This night we camped in [the] Swamp of Marsh Creek, after fording it, about 10 miles from our last Camp.

On the 28th we reached the Bend of the Port Neuf River and Camped near it. The River was swollen by the late heavy rains so as to fill its banks at this point.

On the morning of the 29th we found some of our cattle missing, which had strayed off during the night owing to the inattention to the guard. While waiting for Jacobs and some of the men who had gone out to hunt them, to bring them into Camp, we saw a band of 12 Indians, coming up from down the River on horse back. They came towards us as rapidly as their horses could gallop, and when within half a dozen rods of the Camp, suddenly stopped. This is a usual Custom among peacible Indians. We made motions for them to come into camp, and gave them something to eat. Jacobs soon arriving with Frank, the Frenchman, they told them our Cattle were down the creek near their Wakaups [wickiups], and soon the other boys were seen driving them up from that direction. We traded a few charges of powder and Caps with them for some Deer skins, which we much needed for our ox whips.

In this party was old Snag, a petty Chief of the Snakes & Bannacks, and his son, by whom Sam Batchelder & myself

---

19. Purple here meant the Hudspeth Cutoff, which his party reached in the vicinity of Lava Hot Springs. The Sublette Cutoff was located farther east, between South Pass and Soda Springs. Gregory M. Franzwa, *Maps of the Oregon Trail* (Gerald, Missouri: Patrice Press, 1982), 172-73.

explained that we had lived in California, and were seeking gold diggins in the Country north of us. I have always been of the opinion that this old Indian knew of the existance of Gold on Grasshopper Creek. He marked out on a piece of paper to Sam the locality where gold could be found, which Sam and I comparing in December following, found to correspond exactly with the situation of Bannack City.[20]

We travelled 6 miles to day, having to repair and ascend Cadys hill. Our train was passed by a number of men and waggons, bound like ourselves to the Salmon River Gold Mines. We camped on Cady Creek Hill.[21]

Parties passing us to day commenced calling our train the Mormon Train, because in answer to their inquiries, we told them we had outfitted at Salt Lake.

At noon the next day, 30th June, we camped a couple hours on the creek at the foot of Cady Hill. We observed a number of Indian Wakaups [wickiups], across the little valley through which the Creeks run, and one or two Indian women, gathering berries and digging roots. Shortly after making our noon camp and eating our dinner, the men strolled off, some for the purpose of gathering strawberries, some fishing, while some at the distance of a couple hundred yards, in a shady place by the creek, were washing their shirts, a few remaining in camp to watch it. I lay down in the shade

---

20. Snag or Old Snag was the white name commonly used for Tio-vandu-ah, a Shoshone subchief. Baptized by the Mormons at Fort Limhi (later renamed Lemhi) in November of 1855, he remained generally an exponent of peace and friendship with all whites until his murder at Bannack City in 1863, an unprovoked killing recorded by Purple and others. He and his people were quite familiar with the country around Grasshopper Creek. See Brigham D. Madsen, *The Bannock of Idaho* (Caldwell: Caxton Printers, 1958), 92, 94; and Virginia Cole Trenholm and Maurine Carley, *The Shoshonis: Sentinels of the Rockies* (Norman: University of Oklahoma Press, 1964), 150, 155.

21. At this point the party was moving eastward along the Hudspeth Cutoff, crossing the Fish Creek Range that separates the Portneuf River watershed from the Bear River watershed.

of one of the waggons, smoking my pipe and dreaming of the memories evoked by completing my 31st birthday, and in a few minutes fell fast asleep. I was shortly awakened by the whispered mutterings of voices, and opening my eyes, saw three young Indians standing near me, and a fourth kneeling on the Buffalo Robe on which I lay, was reaching out his hand for my pipe which had fallen from my lips while asleep. I raised myself up, on my as-coxygis, and the Indian with a grunt of apparent satisfaction at having awakened me, pointed to my pipe, and by motions made me understand that he was desirous of enjoying the luxury of a smoke. Filling up the Pipe and lighting it, I gave it to him and his Companions, during which operation I was much surprised to find, upon looking about, that there was not a single one of our men about the Camp. I regarded this as the most car[e]less inattention of which our party had been guilty, during our travel.

That Indians should have come into our Camp without our knowledge, and in Broad day light, annoyed me excessively, although they were friendly, and well disposed in their actions; for had the contrary been the case, we might have repented too late our carelessness.

We moved our Train a short distance further in the afternoon, Camping in the Valley at the Forks of the Road near Bear River, having made Eleven miles.[22]

On the 1st of July we reached a fine spring, on the upper or north eastern side of the road, which came bubbling and

---

22. Coming eastward along the Hudspeth Cutoff, the party had reached Bear River and the intersection with the main Oregon-California Trail route at Sheep Rock (Soda Point), a few miles west of Soda Springs. Here they turned northwest, taking the original Oregon-California Trail's main line toward Fort Hall on the Snake River. This early in the season, of course, and particularly with the disturbed condition of affairs in Missouri and the threat of Indian hostilities west of Fort Kearny, they did not meet any parties of overland travelers coming from the east, but hopeful miners from Colorado were already on the road.

gurgling out at the foot of the hill, about seven miles from our last Camp, where we encamped for the night. We passed through a remarkable looking country yesterday afternoon and to day. Portions of it here seemed to have been visited at some remote period with a terrible up heaving of Nature. Rocks are scattered about the level plane, and frequently we passed over spots, where for twenty or thirty yards, the earth seemed to be hollow, and the noises of our animals feet would make a hollow rumbling sound, as though we were travelling over a Cavern. Near the Banks of Bear River I visited several holes or fissures in the rocks, which were so deep, that I could not hear the large stones which I tumbled into them, strike the bottom, nor plash in the water—if there was water in them. They appeared to be dry, but the truth of this I could not, from their great depth, determine. Jacobs told me [that we] were within five or six miles of the Soda and Marvellous Steamboat Springs.[23]

July 2nd. We camped on, after crossing, the Porte Neuf, which at this point was a mere brook, and gave us no trouble to ford. We travelled about 15 miles to day.

---

23.   Soda Springs and Steamboat Spring were famous points of interest along the Oregon-California Trail. In the town of Soda Springs, Idaho, one of the soda springs continues to produce at a commercial health facility. Most of the rest, including Steamboat Spring, are now submerged beneath the Soda Point Reservoir west of town. The curious can see bubbling evidence that Steamboat Spring is yet puffing away beneath the reservoir waters. Gregory M. Franzwa, *The Oregon Trail Revisited*, 4th ed. (St. Louis: Patrice Press, 1988), 294-98.

24.   Lander's Road was an alternative Oregon-California Trail route, part of the Central Pacific Wagon Road project constructed in 1858 by a Department of the Interior engineering corps led by Frederick W. Lander. Beginning at South Pass, the road led across the headwaters of the Big Sandy River and the Green River, across the Salt River Range, north through the upper Salt River valley for twenty-one miles, thence northwest and west to the Ross Fork of the Snake River (where Purple's party intersected it), and followed that stream to Fort Hall. Beyond Fort Hall the route ran along the south side of the Snake River valley to the Raft River, then turned southwest up that stream to end at the City of Rocks. Despite vigorous promotional efforts by Lander in 1859 and 1860, this route never

July 3rd. We reached the junction of Lander's Road and camped for the night, having made 14 miles.[24]

July 4th. I had by this time become somewhat doubtful of the capacity and ability of Jacobs, as a competent guide. We had since we left Deep Creek, been travelling most of the time, over hills, and through valleys where no track by a waggon had ever been made, before ours. The object we had in view for diverging from the old paths of travel, had been to avoid the crossing of streams, that would have necessatated our unloading and Ferrying our waggons. The heavy rains of the season had rendered the ground in the low valleys so soft, that our heavy waggons would sink through the crust, and we were compelled to keep as much as possible on the ridges and hills, where the ground was firmer. Upon resuming our journey to day, Jacobs had taken us off the Lander Road for a short distance, as he said to shorten the days travel. I had come to the Conclusion that "the longest way round was the shortest way home," and told him that hereafter there must be no diverging from the usual lines, or old routes of travel, and coming into the Lander Road again just after noon, we followed it till we made our Night Camp, near the Head of Ross' Forks, at a point where grew a few stunted Cedars on the Hill side. About noon we were visited with a severe hail storm, which ended in a regular Snow squall, and wet and cold, we were glad to make an early Camp. The Cedars enabled us to make up a good fire, and by night we had dried our Clothing, which rendered us warm and comfortable. We had travelled about 14 miles to day.

---

attracted a heavy emigrant traffic—in part because of Shoshone hostilities. For an excellent account of Lander's road-building activities, see W. Turrentine Jackson, *Wagon Roads West: A Study of Federal Road Surveys and Construction in the Trans-Mississippi West, 1846–1869* (Berkeley: University of California Press, 1952), 191-96, 206-14.

July 5th. I was detained this morning by my Poney, who contrary to his usual habit, had strayed off during the night from the other animals, and I feared had been stolen. The Train moved on, and I remained behind to hunt him up. After looking a couple hours or more, I gave up in despair of ever seeing him again, feeling convinced that he had been stolen. While travelling on the high points of land, where I could have a more extended view of the valleys, and fine grassy plots below me, where if feeding, my Poney would most likely be, I had a most charming view of the valley of Snake River. At the distance of twenty four or five miles, I could discover the River, like a thread of silver, curving and winding its way through the valley, until lost to sight in the western horizen. The three Butes that rise from the level plain of the valley beyond the River, like immense Ant hills, were plainly discernable, but covered with a hazy vail covering, and almost shutting them out from the sight.[25]

While gazing upon the scene, I hear the whinney of a horse, and looking to my right saw my Poney, who apparently had been to Camp, and finding his companions gone, was running around briskly, seemingly much distressed at their absence. Calling to him, he came up gently as a dog to me, and throwing a halter on his head, [I] mounted him bare backed, and in a short time we overtook the train.

We camped to day down Ross Fork, about 16 miles from our last camp. Ross Fork creek is a very bad camping place, down as low as this point. The stream has made [its] way through the low soft ground of the bottom, so that its banks

---

25. The three buttes are Twin Buttes, thirty-two miles west of Idaho Falls, and Big Southern Butte, nineteen miles north-northwest of Twin Buttes. Well-known landmarks along the Oregon-California Trail, the three buttes provided a sight that cheered and inspired many travelers. Purple provides a more detailed description of the buttes below, under date of July 13, after crossing the Snake River. For northbound travelers, these volcanic formations remain in view nearly all the way to the foot of the Bitterroot Range.

are abrupt, and steep, being ten or twelve feet high, and it is almost impassible, for stock to get down to water. We came near losing five or six head of our oxen, who being very thirsty when unyoked rushed for the creek, and before they could be headed off, plunged down its steep banks, from which we had great difficulty in getting them up and rescuing them from drowning.

July 6th. To day we reached Snake River, and camped near the Ferry, which Jack [Jake] Meeks had just established, a short distance above old Fort Hall, having travelled about 18 miles from our last camp.[26]

Our route across the bottom, from our last camp on Ross Fork, was through a luxuriant growth of Sage Brush, four & five feet in height. The Road was dusty and the day hot and sultry, and our poor cattle, enveloped in the clouds of dust that nearly stifled them, with lolling tongues bowed their necks with increasing energy, in their yokes, as they approached the River, in which they soon quenched their raging thirst.

We found at the Ferry some twenty or thirty waggons awaiting an opportunity to cross. The Ferry having been just established by Jack Meeks and Harry Richards [Rickards], was not yet in good working order. Finding that we could not cross at once, we Corraled our waggons, and awaited with patience our turn to go over.

---

26. This account by Purple provides the most specific information we have about the establishment of the Snake River ferry by Jacob Meek and Harry Rickard in 1862. By the following travel season, according to the most reliable emigrant guidebook, Meek had turned over the enterprise entirely to Rickard. J. L. Campbell, *Idaho: Six Months in the New Gold Diggings. The Emigrant's Guide Overland* (New York: J. L. Campbell, 1864), 12.

Cinching packsaddles was a typical start to a day on the trail en route to the Montana goldfields from Colorado's Pike's Peak, Salt Lake City, or other points in the 1860s. (Nathaniel P. Langford, *Vigilante Days and Ways: The Pioneers of the Rockies*, vol. 1, 1890, opp. title page)

CHAPTER TWO

⚜

The Snake River Ferry to Fort Limhi
July 6 to July 20, 1862

Each day brought fresh arrivals of men and Trains, most of whom were from Pike's Peak, Colorado. I presume that during the seven days we were at the Snake River Ferry, there was at least a thousand to 1500 men crossed the River bound for the Salmon River Mines, and many who did not cross went down into the Boise country.

I had heard it reported at Salt Lake City that the Mormons who had settled at Fort Limhi, and were driven from their settlement in 1857 [1858] by the Indians, had discovered Gold, on Limhi creek in large quantities; but upon apprising Brigham Young of the fact, he had told the Discoverer to stop prosecuting his search for Gold, and leave it alone. Brigham always told his people to develop the agricultural products of the earth, and leave Gold & silver where God had placed it.[27]

---

27. At the behest of Brigham Young, a small group of Latter-day Saints established the Salmon River Mission in 1855 on the Limhi (later renamed Lemhi) River, a tributary of the Salmon River in east central Idaho, close to the Continental Divide. Extremely isolated but hoping to convert the Shoshone and Bannock peoples who made that country their home territory, the colonists built a fortified central post they called Fort Limhi, located about two miles north of the modern town of Tendoy. Although President Young paid the colonists a visit in 1857 and praised them for their successes, a short growing season and grasshopper invasions made the colony's existence precarious. In March 1858, while church leaders at Salt Lake City were preoccupied with the invasion of

It was our intention, when we started on our journey, to go to the Mines of Florence or Elk city, then in Oregon. A month prior to our leaving Salt Lake, some friends whom I had become acquainted with during my stay there, and who like myself were gentiles, had left for the Salmon River Mines, and expected to find a route by which they could reach Florence via Fort Limhi, to which place they directed their journey. The names of this party were Jack Mendenhall, Thomas Pitt, H. F. Morrell, Robert Meneffee, Thomas an Irish boy, and two or three others whose names I have forgotten.[28]

Having been told by Jacobs that it would be impossible to get through from Limhi to Florence with our waggons, I had determined, after consulting with my partner Tisdale, to take the Fort Limhi road, when we reached it, and go to that point, if in the meantime we should get no news from Mendenhall & his party. And if we found Jacob's statement correct, we would go over to Deer Lodge. Jacobs told us that the Stuart Brothers were digging Gold in Deer Lodge valley, and advised us to go there, although he could give no definite account of what they were doing in their mining operations.[29]

Utah by federal troops under General A. S. Johnson, an attack by Shoshones and Bannocks resulted in the death of two colonists and the loss of many cattle and horses. When a relief expedition arrived from Salt Lake City, the Mormon colonists entirely abandoned the project and returned to less exposed settlements in northern Utah. Despite the rumors heard by Ed Purple, there is no evidence that the Fort Limhi colonists had discovered any gold, nor were gold strikes later made in this locality. See M. D. Beal, *A History of Southeastern Idaho* (Caldwell, Idaho: Caxton Printers, 1942), 136-52; Milton R. Hunter, *Brigham Young, the Colonizer*, 4th ed. (Santa Barbara: Peregrine Smith, 1973), 357-64; and Madsen, *Bannock of Idaho*, 95-108.

28. All four of the men named by Purple—Mendenhall, Morrell, Meneffee, and Pitt—lived in Bannack City during the winter of 1862–1863, according to the Purple and Sanders census listing: *Contributions*, 1:339-40. For biographical sketches of Mendenhall and Meneffee, including a description of their trip to Fort Limhi, see Thrapp, *Encyclopedia of Frontier Biography*, 2:972. A sketch of the career of Mendenhall, a native of Indiana who later established a prosperous grocery business at Bozeman, can also be found in Leeson, *History of Montana*, 1145.

Meeks and Richards in establishing their Ferry had laboured under great difficulties to obtain proper materials for that purpose. Their Boat had been built over the Porte Neuf and hauled over to Snake River in waggons. Their Cable rope, stretched across the river, which at this point, owing to heavy rains, was fully 250 yards wide between its banks, was brought up from Salt Lake. Their blocks and Pulleys for attaching the boat to the Cable, were Manufactured on the ground, and as they rapidly wore out, gave employment to numbers of Mechanics among the passing Emigrants. [They] Being short of Rope to gear the Ferry Boat to the Cable, I sold them a large coil of inch rope I had brought along, for the purpose of using Myself for crossing Rivers. I also sold them a small Row Boat with which I had provided myself for the same purpose. I realized about $280 for what had cost me $40, and the purchasers were well contented with their bargain.

During our stay at Snake River, a number of Snake and Bannack Indians pitched their Lodges near the Ferry, and were employed by the emigrants in swimming over their horses and mules. These Indians are expert swimmers, and would take to the water as fearlessly as a New Foundland Dog, keeping just behind the stock, sometimes catching hold

---

29. James and Granville Stuart are the best known of Montana's early non-Indian settlers, in large part because of Granville's writings and the journal the Stuart brothers kept. Born in Virginia and raised in Iowa, the two men had gone to California with their father in 1852, then arrived in the Beaverhead valley with their cousin Rezin (Reece) Anderson in late 1857, deterred by sickness from returning east. In spring 1858, accompanied by Anderson and Tom Adams, they successfully prospected for gold in Benetsee or Gold Creek north of Deer Lodge, but then lacked the tools to pursue their discoveries. With reinforcements they resumed mining early in 1862. By July a sizeable emigration from Missouri and Colorado had begun to reach their little settlement, although not all the newcomers remained. See Stuart, *Forty Years*, esp. 1:137-40, 205-14; Thrapp, *Encyclopedia of Frontier Biography*, 3:1379-82; and P. Richard Metcalf, "Granville Stuart," in Howard R. Lamar, ed., *The Reader's Encyclopedia of the American West* (New York: Thomas Y. Crowell Co., 1977), 1147-48.

of their tails, and splashing water on the side of their faces to keep the Mules & Horses in the right direction, and invariably piloting them safely to the opposite side.

On the 9th of July a Horse race came off, between an Indian's Horse and one owned by one of the emigrants. It was a race of 100 yards for a purse of $20, and came near leading to a difficulty with the Indians. The Indian got the start [by] a few yards, and was ahead at the out come. By the Indian rule governing races, he had won. But the judges decided that as he did not come out further ahead than the distance he [had] started ahead of the white man's Horse, he had lost the race.

The money was paid to the white man, but the Indians managed to steal, within three weeks and 200 miles away from the Ferry, the winning mare from her owner.

Just before our arrival a man by the name of Holmes had been killed at the Ferry, by his partner with whom [he] had quarreled in reference to a division of the outfit they had started with from Colorado. The killing had been justified as an act of self defense, but I heard some of the men declare that it was not warrented by the such circumstances. Whatever may have been the truth of the matter, I could not determine it from the conflicting statements.

On the night of the 10th of July, as I was just getting asleep, I heard the report of several guns and pistols, and the voices of men running down the river, shouting and hallooing at some object which was being borne down its rapid current. It was the body of a young man by the name of Johnson, from Colorado, who with two or three others, had upset in crossing the River in the bed of a waggon. It was not an uncommon thing to use waggon beds in crossing, after being Caulked and pitched for that purpose. The party had overloaded their frail shell, and when in the middle of the stream it dipped water and Commenced sinking. Johnson, telling his comrades, none of whom could swim, to "stick to the waggon

Ed Purple and his party crossed the Snake River on Rickards's ferry (Richard's in Purple's narrative) in 1862, its first year of business when it was still co-owned by Jacob Meek. (Idaho Historical Society, Boise)

bed, boys," plunged into the stream and made for the shore. But the noble self sacrificing fellow had not properly estimated his own strength, and that of the rapid current that remorselessly bore him down, and resisting all his efforts for safety, engulphed him in its deep waters.

Persons went down the stream a number of miles the next day, hoping to find the body to give it burial, but were unsuccessful in their search. In a few days the circumstances were forgotten, save by a few friends, and the tide of emigration daily increasing rolled on, like the Broad River's current, as though no ripple of death had for a moment disturbed its onward flow.

The scenes about the Ferry at this time were enlivening and worthy the pencil of an artist. The fresh arrivals would busy themselves in making inquiries of the country ahead, and the location of the Mines, upon which point there was really little information to be obtained from anybody. Some of them in their great hurry to cross the river, would stimulate their companions to promptness and dispatch in getting over, by saying that "it would never do for a company of Pike Peakers to camp for two nights on the same side of a stream."

Our Train had been dubbed the Mormon Train, an imputation which clung to us with some till late in the fall. This had arisen from the fact of our outfitting in Salt Lake City. It was borne with as much good nature as we could command; but one day at the Ferry, my partner Tisdale, a native of South Carolina, determined that whatever name might be given the Train, he would no longer be called a Mormon.

An innocent looking Pilgrim, who no doubt believed us all to be Elders and lesser luminaries of the Mormon Church, accosted Tisdale for the purpose of posting himself about Mormon religious matters. The cup of Tisdale's indignation was now full and overflowing, and pulling out his six shooter

[he] threatened the fellow that should he ever presume to call him, or intimate that he was a Mormon, he would blow the top of his head off and I believe he would have done it.

This little episode, leading to some explanations in regard to who we were and where our party hailed from, helped to place us upon a better footing in the good opinions of our fellow travellers. It may be remarked that there is as strong a prejudice, almost, against the Mormons by persons who have been much on the plains, as there is against the Indians. It arises doubtless from their clanish habits, and the frequent occurence of such circumstances, when they are about, as gave rise to the old grudge between them and the Missourians.[30]

On the 10th, 11th, & 12th of July we were busy getting our stock across the River, and Ferrying over our waggons and provisions. We were compelled to make our Horses & Cattle swim, as the Ferry Boat was too small to transport them. By going a short distance up the river, above the point opposite where we wanted our Cattle to land, we would drive in 30 or 40 head of oxen at a time, and by shouting and throwing sticks and stones at them, get them started towards the opposite shore, upon which they all landed safely, about a mile below.

John Tripp & his son, although Mormons, were first rate men, and had both been brought up as Pilots in Falmouth

---

30. Purple here refers to the armed conflict between Mormon pioneers and non-Mormon settlers in frontier Missouri during the 1830s, which concluded with the Latter-day Saints fleeing from the state after Governor Lilburn W. Boggs issued against them his notorious "Exterminating Order" to the state militia in October 1838. These events receive detailed attention in Max H. Parkin, "A History of the Latter-day Saints in Clay County, Missouri, from 1833 to 1837" (doctoral diss., Brigham Young University, 1976); and Leland H. Gentry, "A History of the Latter-day Saints in Northern Missouri from 1836 to 1839" (doctoral diss., Brigham Young University, 1965). For a thorough, scholarly analysis of anti-Mormon rhetoric and the underlying sources of anti-Mormon sentiment during this era, see Leonard J. Arrington and Davis Bitton, *The Mormon Experience: A History of the Latter-day Saints* (New York: Alfred A. Knopf, 1979), 44-64.

harbour, England. Their skillful management of our small boat, the use of which I retained till our things were all over, was of great service to us. The current of the River was rapid and strong, and we made some thirty crossings, before we landed all our plunder on the opposite shore.

Having bid good bye to Meeks & Richards, and the Ferry hands, we rolled out on the Morning of the 13th of July up the Snake River bottom.[31] This bottom is covered with a thick growth of Sage brush, which covers the level plane, that stretches north west from the River to the distance of sixty or seventy miles before reaching the Mountains, with an occasional sandy patch of ground upon which there is no brush, grass or growing herb of any Kind.

From this plain, & perhaps 35 to 40 miles from the River, rise three Buttes, which are certainly most remarkable developments of nature. The two smaller ones I should judge are about 25 to 30 miles apart, and the larger one, some 40 or more miles from the small one nearest the River.

The small ones are nearly alike in height and size, being I should judge 10 or 12 hundred feet in height, and two miles or more across their base. Their appearance at a distance

---

31. Establishment of the route north from Fort Hall is summarized in Betty M. Madsen and Brigham D. Madsen, *North to Montana! Jehus, Bullwhackers, and Mule Skinners on the Montana Trail* (Salt Lake City: University of Utah Press, 1980), 7-22. Captain John Mullan remarked in 1865 that this road "has been used by wagons for many years, and though not worked is quite practicable." *Miners and Travelers' Guide to Oregon, Washington, Idaho, Montana, Wyoming, and Colorado* (1865; reprint, New York: Arno Press, 1973), 10. Although Purple apparently had no maps, Captain A. Pleasonton's 1850 "Map of the Country between the Cascade and Rocky Mountain" clearly marked the general line of the route. With refinements in detail, the 1859 Bureau of Topographical Engineers' "Map of the State of Oregon and Washington Territory" also indicated the wagon route distinctly. Both are reproduced in *Maps of Early Idaho* (Corvallis, Oregon: Western Guide Publishers, 1972). See also Carl I. Wheat, *Mapping the Transmississippi West* (5 vols. in 6, San Francisco: Institute of Historical Cartography, 1957–1963), 3:128-29, 4:144.

resembles a huge dark rock, of a regular oval form and destitute of vegetation. I can compare them to nothing I have ever seen, save immense Ant hills, which in the regularity of their form, they resemble—but unlike them in their Rocky appearance.

The larger Butte, which lays south & East of the smaller ones, is less simmetrical in appearance, being exceedingly rough & rugged, and I should judge 1500 to 1800 ft. in height and 4 or 5 miles [or] perhaps more across its base. I observed on the 6th of July, 1862, that its top was covered with snow, while the small ones were entirely destitute of this wintry mantle. There are springs of fresh water upon it, and good grazing for cattle at its base.

It is difficult to determine at so great a distance, on a wide extended plane like Snake River Bottom, with much accuracy, the height and size of such Land Marks as these Three Buttes. I have passed them four times during my travels in this Country, and I have never heard of anyone visiting or exploring any but the Larger one. From my friend Mr. C. M. Davis, who visited this one, I obtained the information that there were several springs upon it. I am of the opinion that at some former period of time the places or points where these Buttes now stand were the openings and outlets of hot springs of Mineral water, similar in Character to those found in the Salt Lake basin, and in Deer Lodge Valley; that the water discharging itself from these openings and throwing up mineral matter, mud, and earth from great depths beneath the surface, formed, after many centuries in undergoing this process, these immense hills. That in the course of time, the waters found new channels, or sinking below their old courses, have found new vents for their outlets, nearer the surface of the ocean, perhaps many thousand miles away. The Hot spring mound in Deer Lodge Vally is to day undergoing such a process of gradual formation, and it may as ages pass over its head, lose its waters after having assumed the gigantic proportions of these Buttes, compared with which at present it is a mere pigmy.

The reason why these two small Buttes have not been explored by the passing travellers, has been owing to their distance from the River, and the difficulty of appr[o]aching them, by reason of the close and compact growth of sage brush, through which it is almost impossible to ride a horse, & the deep fissures which abounds in the Valley. A party visiting them, would have to pack water for himself and animal, which would necessarily render his stay in Exploring them very brief. It think it would take a week to examine one of them thoroughly.

When passing along in the sight of the Buttes I asked Jacobs how many years he had been living in the Mountains, and at what time he first saw these Buttes. Like all the old Mountain men I ever met, he prided himself upon his experiance, estimated by the number of years he had been living and roaming about with the Indians, which was really about ten, a fact which he had inadvertantly let slip by once telling me that he came on the plains in 1852, bound for California. In answer to my question this time, he told me, with an air of seeming candour, that when he first saw Snake River, "it was so many years ago, that it was merely a small brook," and "the 3 Buttes were holes in the ground", and that "both had grown to their present proportions since that time." This account afforded our party much amusement, in which Jacobs himself participated, and I related the incident as a sample of some of the Munchausen stories of old Mountaineers.

We camped to night up the River about 19 miles from the Ferry.

On the 14th we got an early start, and continued our journey making our camp for the night near a cluster of Cedars, that grew on the bank of a slough, partly filled with water, near the River. We travelled about 14 miles to day.

July 15th. We were in sight of the three Tretons [Tetons], or Teats, all day. They lay to our right, at the distance of 80 or 100 miles, and are sharp points or Peaks that rise from the Wind River Range of Mountains. They were covered with snow, rough and ragged in their outline, and present a scene of lonely and solitary grandieur seldom met with, even in the Rocky Mountains.

We camped at a slough off from the River, having made about 20 miles.

July 16th. Upon leaving our Camp this morning we passed through, for two miles, a rough and rocky country. We killed a number of Rabbits to day, among the Rocks, which gave us a capital supper. The evidences of volcanic origin are abundant in this section of the country. We passed Market Lake, a small pond of shallow water in which there was a large number of wild ducks.[32] I was informed by Jacobs and Frank that the name Market was derived from the fact that many years ago, Buffalo and other game was abundent in this vicinity. It has now entirely disappeared, and a few Buffalo skulls lying about are the only indications that these animals, ever roamed over this now barren (save of Sage Brush) and sterile Country.

---

32. Market Lake Slough, the remnant of the Market Lake of Purple's time, is directly adjacent to old Highway 91, paralleling Interstate 15, between Roberts and Sage Junction, Idaho. Before reaching Market Lake, the route skirts Hell's Half Acre Lava Field, the southeastern salient of an extensive lava formation that can be seen in its most spectacular form approximately eighty miles northwest at the Craters of the Moon National Monument.

33. From its headwaters in the Targhee National Forest to the northeast, Camas Creek drains into an area of sinks and sloughs set off today as the Camas National Wildlife Refuge near the town of Hamer, Idaho. By 1864, according to Kate Dunlap, it had become customary for emigrant trains to cross the waterless, grassless, sandy country between Market Lake and Camas Creek at night, then travel a second night to reach Beaver Creek at the base of the Beaverhead Mountains. *The Montana Gold Rush Diary of Kate Dunlap*, ed. J. Lyman Tyler (Denver and Salt Lake City: Fred A. Rosenstock Old West Publishing Co. and the University of Utah Press, 1969), B-41.

We camped at Camus Creek, having made about 25 miles, our course having been a little west of north from Snake River.[33] We lay by at Camus on the 17th July till late in the afternoon, having concluded to make our journey to Painted Rock Creek (called Mormon Fork on Delacys Map. This was also called Woods Creek by Jacobs) in the night, to avoid the heat and dust of the day, which was almost intolerable for man and beast.[34]

We reached Medicine Lodge Creek about sundown, and crossing it a short distance above where it sinks, followed the well worn tracks, which had been made by the emigration that preceeded us, who with mule and Horse teams, were passing us every day, after we left Snake River Ferry.

From Medicine Lodge Creek, I went on ahead alone to pick out a suitable camping place for the train on its arrival in the Morning at Painted Rock creek. During the night I met a party of three men, with a waggon, returning from above, who told me that there was great excitement among the people at Fort Limhi, because they could find no Gold in that vicinity, and this party having satisfied themselves of that fact, were retracing their steps to Snake River Ferry, on their way back to the Pike Peaks Country. I paid no attention to their report, because, whether true or not, I did not deem it prudent to go back and turn the train around on that dismal prairie bottom at night, and I knew my Partner Tisdale, with whom I had left it, would not turn back for any stories these men could tell him.

---

34. At this point the Purple train had turned northwest, taking the route that headed toward Fort Limhi. The wagon road toward the Beaverhead valley and Deer Lodge, to which Purple would return after his Fort Limhi excursion, led from Camas Creek almost directly north toward Pleasant valley and Monida Pass. For the De Lacy map, a map of Montana Territory published in New York City in 1865, see Wheat, *Mapping the Transmississippi West*, 5, part 1:148-53. The Mormon Fork shown by De Lacy corresponds with the Lemhi River, but to reach its headwaters Purple needed to travel from Camas Creek to Medicine Lodge Creek and Crooked Creek—the stream he called Painted Rock Creek.

I reached Painted Rock creek long before daylight, and just after sunrise on the 18th July the train arrived, having travelled 28 miles since leaving Camus Creek.

We found, camped at this point, a large number of persons who were waiting to hear from Fort Limhi, which was about one hundred miles above us. The rumor was that the advanced emigration that had reached that place, had been unable to find any diggins, and were breaking up their waggons, from the hard tough wood of which they were constructing Pack saddles for their Horses & Mules, and intending to Pack over the Mountains to the Gold Diggins at Florence and Elk City. It was also rumored that many of them were greatly exasperated at the d—md Mormons, as they called them, for spreading a report of the existance of Gold in that quarter, by which they had been so grossly deceived, and a long tedious, and useless journey imposed upon them. It was intimated that this feeling was so strong with some of them, that they were ready to punish in the most summary manner, by hanging by the neck, any Mormons they could catch, regarding them as they did, to be the authors of all their disappointment and distress.

I regarded these rumors, if true, to be the wild and distempered phrensy of an unreasonable, disappointed set, who having set out upon a Gold Hunting Expedition with only a vague idea of the locality they sought, and of their ultimate destination, were seeking to shift the responsibility of their own actions, in making such a journey upon imperfect and meagre information, to the shoulders of other and innocent parties.

As I could find no one who had been to Fort Limhi, all of those returning having turned back upon the strength of the stories that were told before reaching that point, I determined to go there myself, and ascertain the exact condition of affairs at that place.

From this project my friend Samuel W. Batchelder tried very hard to dissuade me. The love between us for many years had been as warm and pure as that of Jonathan and David; he feared the excitement might culminate in violence to my person, the danger of which would be more imminent by being away, alone, and so far from the protection of my friends, should an attempt by made to carry out the rumored threat.

I was not insensible to the weight of his argument, but could not help believing that it was my duty, first to relieve our party from the imputation of being Mormons, that had so unjustly obtained credence with our fellow travellers, and at the same time obtain for ourselves such information as I could of Mendenhall & his party, as the only way we could intelligently determine our future course of travel.

On the afternoon of the 18th in Company with Geo. Dewes, a young Kentuckian, we started to Fort Limhi. Our mode of travel was to camp about sun down, eat our supper, and then travel till 10 or 11 o'clock at night. By early day light we would push on till 8 or 9 o'clock, then take Breakfast, lay by till about 4 o'clock in the afternoon, move on till sundown, take supper and resume again our journey a few hours in the night. In this way we were never near the smoke of our Camp fires but for a short time, and less likely therefore to be discovered by prowling parties of Indians, who might feel disposed to int[er]fere with us.

As we were making our afternoon ride on the 19th July, we discovered two men in the trail ahead of, and travelling towards us. We thought at first they were Indians, but upon rising [over] one of the Knolls we acertained they were Whitemen, coming from Fort Limhi.

They proved to be a Mr. Goodel, a gentleman from Ohio, and an Irishman by the name of Lily, a brother of the noted Prize Fighter of that name (Christopher Lily).

Before they had come up to us, George and I had concluded that whether friends or foes, we were evenly matched
and we would have a parley with them.

"Where is Purple and His Train?" was the first salutation
that greeted our ears, from the lips of the Irishman Lily. I
replied that is my name Sir, and my train is camped forty five
or fifty miles below here, on the creek. "Where is that Son of
a bitch, Jacobs? who has inveigled us here into a niche of the
mountains," says the Irishman, in the most insolent manner
he could assume, "to starve to death, or buy his flour at such
thieving prices, as he may ask." It is well for you that Jacobs is
not here, or he would make you eat your words, I replied; you
have no right to talk in that manner of a man of whom you
know nothing. Jacobs, I continued, was hired by me, to guide
my Train, and is responsible to no body in this world but
myself and Partner for so doing. And as for being a Flour
trader, he has not got 300 lbs. of that article with him.

"Oh never mind" says Mr. Goodel to me, "Never mind
what he says," meaning the Irishman, "he is excited," and
speaking in a low whisper to me, "he is a d—md fool beside."
They then told me that they had been appointed a Committee by a large number of the men at Fort Limhi, to visit my
train and ascertain what information they could, in regard to
Gold Diggins in that locality, and if I had no knowledge of
any, to find out the proposed destination of my Train.

I told them what rumors I had heard of Gold Diggins at
Limhi, in which however I had little faith; that the object of
by visit to Fort Limhi was to find the whereabouts of my
friend Mendenhall & party, and that I should be governed by
the Report I might have from them in regard to my ultimate
destination.

In half an hours time we had established some confidence
in each other, and Goodel & Lily turned back, to accompanie
us to Fort Limhi. We made our Camp about sun down, as

usual, and George and myself being better off for provisions than they were, shared our food with [them].

After our supper, by aid of those mystic signs and that language, by which the members of a great brotherhood, foreign in their birth and tongue though they may be, recognize each other, in whatever quarter of the globe or under whatever circumstances they meet, Mr. Goodel and myself became acquainted; not as just before, simply as men or gentlemen to whom the conventional ordinary courtesies of life were to be extended, but as members and Brothers of a noble & glorious Order, whose first teachings, as we stood upon the threshold of the Temple as Neophytes, had been charity and love, peace and good will towards our fellow men. Here in this lonely and secluded place, far from the haunts of men, and the scenes of refined civilization, menaced by an undefined danger, I found a man, rough in his exterior, and in whose breast lurked perhaps a wicked purpose of mischief and injury to fellow being, but now touched by the magic wand of Free Masonry, "the better angel of his nature," subduing the baser passions of his soul. He clasps my hand and bestows the token of faith, which lights with hope and charity, greater than all these—that makes us indicated to be members of a common & almost universal Brotherhood.[35]

I need hardly write here, what may be so well supposed, that this was a most pleasant and agreeable interview, and though I do not doubt, but what I should have been able to allay the excitement of the past few days, which had without reason or judgement, directed itself, towards us—yet I could not fail to be sensible of the potent influences invoked in our

---

35. On the Masonic order in the northern Rockies during this era, see Cornelius Hedges, "Early Masonry in Montana," *Rocky Mountain Magazine*, 1 (September 1900), 13-17; and Llwelyn Link Callaway, *Early Montana Masons and Other Writings of Lew L. Callaway, which have appeared in the Mason Star and Spire* (Billings: Western Litho-Print Press, 1951).

favor, by the knowledge of what I at least was, and of which Mr. Goodel was now possessed.

Making a short ride after our supper, we camped early for the night.

July 20th. Resuming our journey early in the morning, and pushing on as rapidly as compassion for our Animals would dictate, we reached Fort Limhi about 3 o'clock in the afternoon. A few of the old log houses, built by the Mormons, still remained, some in a tolerable habitable condition, to mark the spot of the former settlement. There was probably 4 to 500 people camped there, and many had broken up their waggons, and Packing their animals had started over the Mountains, for the Oregon Mines. As I rode up towards the square formed by the old log houses of the former settlers, I saw a Tent, in front of which [stood] a rude platform of boards torn from the roof of some of the old houses, on which one or two cut glass decanters and black bottles filled with whisky were sitting—and which I had no difficulty in determining was the saloon and Bar of this recent, five day old settlement.

While speaking to the Proprietor, (whom I afterwards became well acquainted with) J. M. Galloway, I saw coming down the hill on horseback from the east my friend Jack Mendenhall, and Fred Burr from Gold Creek.[36] The latter gentleman I was unaquainted with. I was much pleased to

---

36. Fred Burr had come to the northern Rockies in 1853 as an engineer with the Northern Pacific railroad survey led by Governor Isaac I. Stevens. He remained as second-in-command for the Mullan Wagon Road survey, then located in Bitterroot valley and began life as a trader and rancher. In 1856 he brought a herd of four hundred cattle into the country from Salt Lake City, tended by Jim Minesinger and John Powell. The Stuart brothers were his good friends, and Granville married a sister of Burr's Shoshone wife. On July 17 Granville Stuart noted the departure of Mendenhall and Burr from Gold Creek on his return trip to Fort Lemhi. Stuart, *Forty Years*, 1:214; Thrapp, *Encyclopedia of Frontier Biography*, 1:196.

BOZEMAN.

Ed Purple's friend Jack Mendenhall, pictured here circa 1880, mined at Gold Creek. (M. A. Leeson, *History of Montana, 1739– 1885,* 1885, p. 153)

meet Jack, and after taking a drink, he gave me an account of himself as follows.

He reached Fort Limhi with his party early in June. They found in one of the old houses four men, who [had] come up from the Florence Diggins, on a prospecting tour, and who were entirely destitute of provisions, and nearly so of lead & powder. Their names were Bosdick [Ephraim Bostwick], Jeff Pursell, John Haley, and another whose name I have forgotten.

They had expected to find a settlement of white people at Fort Limhi, had been compelled to leave in the mountains or kill for food their Mule, the only animal they started from Florence with, and finding no game, had subsisted for three days before their arrival at the Fort on a single Sage Hen which Bosdick had killed. They had but two or three charges of Powder upon reaching Limhi, one of which had been used after in killing a chicken, which Bosdick with the instinct of a true hunter had insisted on killing by shooting off its head, while Pursell had pitiously pleaded with him to fire at its body, fearing he might miss it entirely.

Purcell and Haley were found lying on a bed of leaves, in one of the old houses, so weak and emaciated they were hardly able to stand. Bosdick, although much reduced, was stronger, and the only one of the party that was able to get about well on his legs.

The shoes of the party had been worn out, and their feet, cut by the stones and rocks, and pierced by the thorns of the Cactus or prickly pear, were in a terrible condition.

The party would have inevitably starved to death had it not been for the opportune arrival of Mendenhall. They were so weak and their stomachs in such a condition, that they required the most careful preparation of their food, which Mendenhall, Tom Pitt, Morrell, Bob Menneffer & others carefully and kindly provided for them.

While camped here, an attempt was made by the Indians early one morning to steal and drive off some of their Cattle, which would have proved successful had not the watchfulness of a large Mastiff Dog, Ceasar, belonging to Jack, warned them of the danger, and arousing the Camp, the Indians were driven off, after having a few shots fired at them.

Learning from Bosdick the impracticability of attempting to go from that point to Florence, over the Mountains, Mendenhall determined to find his way over to the Bitter Root, or the Deer Lodge Valley, where Burr, Tom Adams, and

---

37. Like Fred Burr, Thomas Adams had reached the northern Rockies as part of the 1853 Stevens railroad survey, then worked on the Mullan survey before settling in the Bitterroot valley and entering the trade with emigrant wagon trains at Fort Hall. He was with the Stuarts and Reece Anderson when they discovered gold at Gold Creek in May 1858. After taking part in the Gold Creek and Bannack City mining excitement, he left Montana and returned east in 1864. For biographical sketches, see W. S. Bell, "Perry W. McAdow and Montana in 1861–1862," *Montana The Magazine of Western History*, 2 (January 1952), 46; and Thrapp, *Encyclopedia of Frontier Biography*, 1:8-9.

some other of his friends, or former gentile acquaintances at Salt Lake city, he knew were leading a Mountain life.[37] He cached, or buried in the ground the bulk of his Provisions, Sugar, Flour, Bacon &c., within the enclosure at the Fort, and packing some of the most gentle of his oxen with a few necessary things for the journey, set out a week or ten days after his arrival at Fort Limhi with his own party & the new found companions for the Deer Lodge Valley.

Taking his course nearly due east, which by the map he had, and although a lately issued one, was incorrect in exhibiting the true course of streams, and mountain ranges, was their only guide, he found himself and party after a week's travel on the Jefferson Fork of the Missouri, but as they supposed on the Bitter Root River. Following it down, nearly to its mouth, they became convinced that they were lost; and finding no game, having become short of provisions, they killed one of their young cattle, and *jerked* the meat. (This is a Spanish term, and means the cutting up of flesh in long thin strips, and drying it in the sun.) A day or two after jerking

---

38. Very likely this Joseph AEneas was the same person known as Aeneas or Young Ignace among the Salish (Flathead) and Kutenai people. His father may have been Ignace La Mousse, otherwise known as Big Ignace, an Iroquois from Caughnawaga, Ontario, who had come west as the leader of a Hudson's Bay Company trapping brigade in the 1820s, lived among the Salish, and taken a Salish or Kutenai wife—perhaps the mother of Young Ignace. It was Big Ignace who undertook the famous journey downriver to St. Louis in 1835 and met with Indian Superintendent William Clark to request that "Black Robe" missionaries come to the Salish country in the Bitterroot valley. Young Ignace or Aeneas was part of a similar expedition in 1839; he and a companion then accompanied Father Pierre-Jean DeSmet back to the Bitterroot valley in 1840. In 1854 he guided for Captain John Mullan; Gustav Sohon, the artist with the Stevens exploring expedition, drew his portrait at this time. An earnest Roman Catholic, Aeneas rose to prominence as a Kutenai chief. He became a strong advocate of peace with the whites and a highly respected leader among the confederated tribes of Salish, Pend d'Oreilles, and Kutenais at their reservation in the Bitterroot valley. See John C. Ewers, "Iroquois Indians in the Far West," *Montana The Magazine of Western History*, 13 (April 1963), 8-9; and Olga Weydemeyer Johnson, *Flathead and Kootenay* (Glendale: Arthur H. Clark Co., 1969), 265-70.

Ed Purple met Sam McLean en route to Deer Lodge. McLean drove a wagon, whatever the terrain, because his size precluded him from riding horseback. (National Archives, Washington, D.C.)

their Beef they were discovered by a small hunting party of Flat Head Indians, of whom Joseph AEneas, a well known Indian of that tribe, was the principal personage.[38] The Indians were much surprised to see a party of white men packing Cattle, like horses, about the country, and making friendly signs they led Mendenhall to their Camp, where he found AEneas, who speaking the English language fluently, for an Indian, was engaged to conduct the party to the residence of Fred Burr on Gold Creek, where they safely arrived in five or six days time. Remaining there a few days, Mendenhall had hired Burr to guide him to Limhi to look after his cache, where I fortunately met him, as before stated.

He was much astonished to find so many people, where only a month before he and his companions had left the place, deserted and desolate, as the howling wilderness that surrounded it. He informed me that a party of men were engaged in mining on Gold Creek, and though not much acquainted with such things, he thought from the appearance of their sluices, when cleaning up, and the pans of earth he had seen washed, that there was Diggins that would pay $2 to $3 per day to the hand, and strongly advised me to [go] over there with my train.

The report of Mendenhall & Burr was eagerly listened to by the men collected there. Goodel introduced me to a number of his friends, who pressed me with their hospitalities, and urged George Dewes and myself to remain with them during the night. Among those with whom I became acquainted here were Sam McLean, afterwards first delegate from Montana to Congress, and Judge H. P. A. Smith, well known in the early history of the Territory.[39]

---

39. Samuel McLean, a native of Pennsylvania born in 1826, moved west to Colorado in 1860, carrying along his legal training and a full load of political ambition. A confirmed Democrat, he served briefly as the attorney general of the provisional Jefferson Territory in 1860. Upon the creation of Montana Territory in 1864, he became the first territorial delegate, serving from 1865 to 1868. Following some success in mining, he later moved to Virginia, where he died in 1877. Kathryn Allamong Jacob and Bruce A. Ragsdale, eds., *Biographical Directory of the American Congress, 1774–1961* (Washington, D.C.: Government Printing Office, 1961), 1310-11. McLean's companion, Judge Henry Percival Adams Smith, born in New Hampshire in 1820, was also a veteran of the Colorado gold rush and Colorado politics. A lawyer and former probate judge in Denver, Smith relocated to Bannack City in 1862, then moved to Virginia City the following year. Although "somewhat demoralized" by alcohol, according to Wilbur Fisk Sanders, he was "a man of very remarkable ability," and he served in Virginia City as a defense lawyer for various accused killers, including two of Henry Plummer's alleged colleagues-in-crime who were hanged by the vigilantes. For his role in defending these men, the vigilantes banished Smith for two years. After a sojourn in Arizona he returned to Helena, where he practiced law until his death in 1868. Thrapp, *Encyclopedia of Frontier Biography*, 3:1327-28; Wilbur Fisk Sanders, "The Story of George Ives," in Helen Fitzgerald Sanders, ed., *X. Biedler: Vigilante* (Norman: University of Oklahoma Press, 1957), 52-53, 78.

CHAPTER THREE

※◆◇◆※

Fort Limhi to Gold Creek
July 20 to August 7, 1862

Having informed Goodel and his friends of my intention to take my train to Deer Lodge, and promising to post up notices at the point where we turned off for Dry Creek, for their guidance as they had also concluded to follow us, I bade them adieu, and with George started back that evening for our Train, and camped a few miles below Fort Limhi, on the creek.

On the 21st July we started early, and made six or seven miles before sun rise. We met today a number of trains on their way up to Limhi, and all were anxious to know the news from that place. A number turned back on hearing our report of the Deer Lodge Diggins, while we left many undecided in regard to the course they would take.

In passing one Train I related the fact of Mendenhall's party *Caching* their provisions &c., and of his returning to Fort Limhi to look after its safety.

"Look a here Stranger," said one of the party, a low, square built man, who made up for his deficiency of height and size, by the important style and manners he assumed, "you say thar's no gold round Limhi, and there was not people thar, when your friend first got thar, and now you tell us he *Cashed* his goods before he left, and came back to raise it. Now how the h—l could he do that. You cant josh me, with that story;

that [is] what an Injun would call talking forked, and it don't go down with this child."

I was amazed to find a man who had surely travelled on the planes as far as from the Missouri River, so ignorant of such a common phrase as "Cache" or "Caching," French words which means burying or hiding, and turning to him, to enlighten him as to the meaning of the terms, I was saved the necessity by the immediate shout of laughter from his comrades, amid which, with their explainations, the fellow slunk back in the crowd, and down to his original diminutive proportions.

July 22. We started early as usual on our journey, and to avoid the Trains comeing up, we kept back on the small trails running parallel to and a short distance from the waggon Road. These Trails leading over small knolls, which were difficult of ascent by the waggons, the Road had been made around to avoid them—while we being on horseback, lessened our travel by following them. We also saved much time by avoiding the trains, for we were compelled to stop and tell the news to each one, and would frequently meet four or five in travelling a single mile.

On this trip we saw a great many Antelope, and several Grizzly Bear tracks, but fortunately, perhaps, none of the animals themselves. In passing down the side of one of the knolls, I heard the rattle of a snake, which George's horse had disturbed as he passed along just ahead of me. He had evidently been sleeping on the warm hill side, and unused to intrusion, was now sounding his notes of war and death.

I always made it a practice to kill these reptiles whenever I came across them, and jumping off my Poney, made for him with rocks and stones leveled at his diamond shaped head. Contrary to what I expected, he showed fight, and not content to act on the defensive by running away, he threw himself into his coil, and leaping towards me two or three times, soon came into an unpleasant, and dangerous proximity to my person. Rocks were scarce, and Poney and I had to beat a retreat to escape his venimous fangs. At a distance I gathered some more stones, and returned to the fight, but his snakeship, no doubt thinking that matters were getting serious with him, darted into his hole in the earth nearby, and thus thwarted my murderous intentions.

There are many of these reptiles in this section, but I believe they are not prolific in the propagating, as they rapidly disappear in the settlements of all new places, where at first, they numerously abounded. It is a singular fact that they burrow in the earth, with owls, and Prairie Dogs, with whom they live on terms of peace and apparent friendship, if such a sentiment can be said to exist in such a low order of Gods creatures. I was told by Col. Sam McLean that in this country he one day came upon a Rocky Den of these snakes, a hundred or more in number, laying about on the Rock, and the exhalation of whose bodies and breath, from their hissing, was so sickening that he was glad to beat retreat, and leave them unmolested.

This creek, which I have called Painted Rock, the name by which Jacobs told me it was known to the old Mountaineers, rises in the Mountains near Limhi. It is not of any considerable size, and grows less as it approaches the Snake River bottom, in which its waters are finally lost or sunk, as are all the small mountain streams which here flow in this direction.

Our Train had been camped a mile or two up the creek, above the point where it sinks, where grew a clump of cotton wood and willow trees, in the vicinity of which there was most excellant grazing for our cattle. For a number of miles above this point, the scenery on the creek was extremely wild and beautiful. On the left the low hills receded a short distance from its pebbly bed, forming a small vally of three or four hundred yards in width, then rising in undulating succession, till they reached, within a short distance, the Salmon River range, whose Peaks covered with perpetual snow, lifted up their towering heads, the seeming sentinals and barriers of the land beyond. On the right, the mountainous appearance of the country was comparatively lost, but enough of the grand and beautiful was to be seen, here, to fill our souls with admiration and reverential awe, towards the great Architect, whose hand had fashioned and formed these wonderful works of nature.

Rising, at the distance of a hundred yards from the creek, and extending up its vally for two or three miles, was a wall, apparently of Red Free stone rock, from 80 to 100 feet in height, whose regular and perpendicular sides presented an insurmountable obstacle to any attempt to scale them. It appeared to have been formed by the action of the pent up waters of the creek, which at some remote period of time may have been damed up by the hills and mountains that environed it, until its increasing and powerful volume had found vent, which had washed down, and cleft in twain the hills, on this side, in its terrific passage.[40]

The immense rocks that are strewn upon the valley below, gives this supposition additional weight for belief.

The continuity of this wall, although varying in its height, was unbroken save in two or three places. In one of these it receded back into the hills, for the distance of a thousand feet, and describing a half circle, returned to its base line, at a distance of six or seven hundred yards from the upper point of recedence, still retaining its abrupt and precipitous character, except in the center of the curve. Here a pillar of rock, with a top surface of fifty by seventy five feet, jutted out, covered with grass, which had sprung up in the soil formed by the falling debris, and elevated at least 60 feet from the plane below. George and I christened this vast and beautiful amphitheatre, the Council Chamber, and it required but little imagination as we sat in its grateful shade, to believe that the voice of a man from the Pulpit rock could have easily been heard by 20,000 spectators, so clear and ringing was our own, as it reverberated from its rocky sides.

I observed on this trip many small banks, formed along the creek by the washing down of the little knolls, in which thousands of our home swallow birds had honey-combed their nests, and which seemed to be the favorite places for their home building.

These little birds, and our common house fly, are about the only reminders we see of home, and it is a remarakable fact that the Fly, puts in an appearance at our Camps, within two or three days at fartherest after its location. I have found them, in my wanderings, in the most secluded spots, and in such isolated places, that I had no doubt the foot of a white man had never before trod. I cannot help thinking but what they are as common and universal in the economy of nature, as are the seeds of her growth and decay.

---

40. This description identifies Purple's route with relative certainty. He and his train had reached Medicine Lodge Creek at the foot of the Bitterroot Range, proceeded in a westerly direction, skirting the mountains, to reach the Crooked Creek watershed, then turned in a north-northwesterly direction along the route of an improved road that is marked on modern USGS maps as Emigrant Trail Road, which turns into the unimproved Crooked Creek Road as it enters the mountains. If one drives west from Dubois, Idaho, along state highway 22, the modern parallel to this route will be reached two-tenths of a mile west of mile marker 51. At this point, avid trail seekers will take the road marked Crooked Creek, following it through stock gates and dodging range cattle, to reach the painted rocks at a distance of 11.7 miles from the highway. The site appears to best advantage early in the morning, as Purple saw it, with the sun shining full on the exposed cliffs. Even under these conditions, however, the skeptical will suspect that distance, time, and the promptings of a romantic imagination had enhanced the scene in Ed Purple's memory. This route will take the determined traveler over a pass to Scott Canyon and upper Birch Creek valley, and thence north-northeast over the Middle Range into the upper Lemhi valley. It saved Purple and others a day's time compared to a possibly less arduous route that would reach lower Birch Creek somewhere west of Scott Butte.

We passed to day a new made grave, which I had not observed on our upward trip. Upon inquiring of some members of a Train, near by, I learned that the young man, a Frenchman, in taking his loaded gun out of the waggon, the day before, had carelessly seized it by the muzzle, and pulling it towards him, the hammer caught in some blankets and falling upon the cap, discharged the contents into his body, killing him almost instantly. The Train to which he was attached stopped for an hour, his friends wrapped the body in blankets, and burried it near the road side, filling the grave with round cobble stones to prevent the wolves from digging it out.

The impressions we have of dying and death—always sad and mournful as they are—are doubly so when brought to our view by such circumstances as these. Insensible indeed must be the feelings of him, who sees cut down in the bloom of youth and the strength and pride of his manhood, a fellow being, far from the home of his kindred and native land, and in this remote and secluded wilderness, finding his last resting place—without feeling a yearning in his heart for a different fate, and believing that

> "On some fond breast the parting soul relies
> "Some pious drops the closing eye requires;
> "For who to dumb Forgetfulness a prey
> "This pleasing anxious being e'er resigned
> "Left the warm precincts of the cheerful day
> "Nor cast one longing ling'ring look behind?"

Saddened with sympathy for the untimely and melancholy fate of this young man, we pushed on our journey, and late in the evening of the 22nd we reached the Train. The boys were much pleased to hear from Jack Mendenhall, and of the better feeling that existed by clearing up the silly reports concerning us. And tired and weary, I sought my bed chamber, canopied by the heavens, and pillowed by mother earth, on which I soundly slept till morning.

July 23d. While [I was] absent to Limhi, Jacobs had left us, having been employed by an old man and his family, to guide them over to the Deer Lodge Valley road, on their way to Walla Walla, Oregon, where some of the Emigration had decided to go. We started out early in the morning, taking an easterly course, and soon struck Jacobs trail, which we followed, reaching our night camp near Medicine Lodge Creek, having travelled about 20 miles from Painted Rock Creek.

July 24. We travelled to Camus Creek, making about 10 miles from our last camp. Here we found encamped a party of Mormon men, among whom was the noted Wm. Hickman, one of Brigham Young's reputed Danites, or Destroying Angels, and two or three of the Luce brothers, of Salt Lake City.[41] One of these brothers, Jayson Luce, a little more than a year afterwards, killed a man in Salt Lake, by the name of [William] Bunton, who had used some insulting language towards him in Bannack City a few months before.

Bunton was himself a reckless character, and being on his way from Idaho to California, passed through Salt [Lake] city, where Luce meeting him on the steps of the hotel, plunged a bowie knife into his breast, and laid him instantly dead at his feet. Although a Mormon, and [a] member of the Church, the wild and dissolute life he had led for a number of years had brought his name into bad repute with the authorities at Salt Lake, by whom he was tried, and convicted of Murder.

By the law of that Territory, a criminal under sentence of death has choice of the means of his execution—either hanging by the neck, or shooting. Luce chose the latter mode, and being led into the jail yard on the day of Execution, was

---

41. On William Hickman, see Hope A. Hilton, *"Wild Bill" Hickman and the Mormon Frontier* (Salt Lake City: Signature Books, 1988).

seated on a low bench, a patch on his breast over his heart to ensure precission of aim, from the half dozen rifles leveled at it from the upper story windows. Everything being in readiness, the signal was given by the sheriff, and five bullets pierced his breast, he toppled from his seat, fell to the ground, and expired without a groan. The sixth rifle was loaded with a blank cartridge, so that the persons acting as Executioners might have the benefit of some doubt as to who really had done the fatal work. This execution took place on my return home, early in the winter of 1863-4.[42]

These persons travelling with horses and mules left us early the next morning, and though they behaved well toward us, I was glad to be rid of their Company.

July 25th. We made a good day's travel today, and camped well up on Dry Creek, having made about 22 miles from our last camp.[43]

July 26th. I met today with a serious accident to one of my waggons. We had not gone more than two or three miles on our journey before we reached a very rough stoney piece of road. I had gone ahead as was my custom, to look out the bad places, and observing the huge boulders in the road, which it was impossible to avoid owing to the narrowness of the defile through which it passed, I stopped and cautioned the men to take time, and ease the waggons as gently as possible over the Rocks. Some of them had passed safely, but when John Murphy's waggon struck it, with that dumbheadiness and lack of judgement so often seen in Irish character, or as I charatably tried to think, from his want of skill in driving

42. On the Bunton killing and the last days of Jason Luce, described here so vividly by Purple, see R. E. Mather and F. E. Boswell, *Vigilante Victims: Montana's 1864 Hanging Spree* (San Jose: History West Publishing Company, 1991), 124-25.

43. Purple's Dry Creek is known today as Beaver Creek, an intermittent stream whose canyon provides a relatively open passage leading toward Monida Pass, the route now taken by Interstate 15.

oxen, he went pounding against the rocks so hard that before
the wheels had made one complete revolution, the lower hind
one had its spokes crushed out and broken off at the hub, and
down it fell, an irrepairable wreck.

I had two tuns of Flour and other provisions on this
waggon, and my other ones were nearly as heavily loaded,
which totally prevented me from thinking of adding more to
their weight. I had some seasoned timber with me, but not
sufficient to fill with spokes the hub of the broken wheel. I
thought I was in as bad a fix as the dirty man on a plain, forty
miles away from water, to wash and clean himself with.

I may say now, without egotism but with a feeling of
pride for the character of the men with whom I travelled, that
they showed their attachment and friendship to me, and set
about at once relieving me from the effects of this serious
calamity. With the exception of Jacobs, Frank the Frenchman
& his Indian woman, and the Mormon Morrisite, his wife &
son, who had left us at Snake River, and Henry Gilbert &
Brother who had gone back from Painted Rock, our party
remained the same as when we left our Camp above the Bear
River crossing. No bickerings or grumblings had marred the
good feeling of each member of the Train towards each other
on our journey, save in one instance to which I shall hereafter
allude.

My orders had been always cheerfully obeyed, and the
duties, for our common safety, which as their nominal head
they had imposed on me, I had endeavoured to discharge in
as mild and conciliatory a manner as the circumstances on all
occasions would admit, with a proper regard for our security.
I believe the boys all liked me, and the feeling was reciprocal.
They piled onto their own waggons, after dividing it up, the
contents of mine, and clearing the Road of the wreck, we
moved on. This accident entailed a loss, beside the trouble
and loss of time, of about $150 on my Partner and self.

We made about 7 miles today and camped at Red Rock Creek, having passed from the waters of the Columbia to those of the Missouri River.[44]

July 27th. We travelled 15 miles today and camped near a beautiful spring some distance from the trail, which Frank the Frenchman told me was called by the Mountain Men Head of Missouri Cold Springs. Such a name might well be given to hundreds of other just such places in the Country round about here.[45]

P. C. Woods killed an Antelop today, great numbers of which can be seen scampering like flocks of sheep over the hills.

July 28th. We camped tonight at the head of Sage Creek, a mere brook which flows I should judge into Red Rock creek, having travelled about 20 miles from our last Camp.

July 29th. At noon today we rested a couple of hours, as we were accustomed to do whenever we travelled 15 or more miles in a day. Our noon camp was at the first crossing of Wolf creek, so called by Frank, and after leaving it I travelled on ahead of the Train a few miles and waited for it to come up, resting myself and Poney in the shade of the Cottonwood trees that grew along the bank of the creek. On its arrival to

---

44. Most likely Purple and his company had crossed the Continental Divide in the area of Monida Pass, then—near the later Williams Junction, a few miles north of the pass—taken a north-northeasterly route to reach and cross the Red Rock River, a tributary of the Beaverhead River, toward the upper end of modern Lima Reservoir. By this route the train would reach the Clover Creek drainage and the headwaters of the West Fork of Blacktail Deer Creek, and then follow Blacktail Deer Creek downstream along a trail route now marked approximately by Montana State Road 202. For this route see the brief description of the Corinne Wagon Road in Frank Harmon Garver, "Early Emigrant Roads and Trails into Montana," typescript, pamphlet 1026, Montana Historical Society (hereafter MHS) Library, Helena.

45. In early August 1864, somewhere near this same point, Kate Dunlap's party nooned at a locale called Johnston's Springs, which she described in these words: "Oh what a nice place! A spring large enough to turn a saw mill bursts up at the base of the mountain. Plenty of good grass." Dunlap, *Montana Gold Rush Diary*, B-41.

where I was, Tisdale told me that one of the oxen had been taken sick, that it had been unyoked and turned out near the noon camp just after starting, and from its actions was by this time probably dead. The supposition was that the ox had been poisoned by some herb it had eaten. I went back to look after him, and found him standing up and feebly feeding on the grass around him. Driving him into the road, I reached the Camp the boys had made at the second Crossing of the creek, about sundown, with my ox. We travelled today about 18 miles.

July 30. My sick ox was able to travel and keep up with the Train without difficulty. I sold him before the close of the year for $85. We made about 20 miles today and camped for the night, near the mouth of Wolf Creek, on the Beaver Head River, having made about 20 miles.

This creek, called by Frank Wolf Creek, is undoubtedly the one named by Lewis & Clarke McNeil Creek.[46] It is called by De Lacy Blacktailed Deer [Creek], and by that name was generally known after the settlement of Bannack City. The Beaver Head is the main stem of the Jefferson River, a fork of the Missouri, which latter name was bestowed upon it by Lewis & Clarke. It is a pity these names are not retained, but the habit of the people in this country of using others has almost obliterated the original ones from their Creeks and Rivers. Big Hole and Stinking Water [Ruby River] were named by these First Explorers respectively Wisdom and Philanthropy Rivers, but these names [if] now applied to these streams would render them unknown probably to a majority of the present inhabitants. The same

---

46. For Meriwether Lewis's account of the naming of McNeal Creek, see Gary E. Moulton, ed., *The Journals of the Lewis & Clark Expedition* (8 vols., Lincoln: University of Nebraska Press, 1986–1993), 5:83-84. This remark and the subsequent comment regarding the De Lacy map demonstrate again that Purple wrote his narrative some time after the event, with the resources of a good library at his command.

may be said of Willard Creek, called now Grasshopper by the party who first dug Gold on it.

July 31. We had now reached a point where we had been determining for the last two days, from the appearance of the country, to stop for a short time and prospect for gold. Wm. Cole, P. C. Woods, Sam Batchelder, and myself were old Californians, having spent some 10 years or more in the Diggins of that state, and in the vernacular of that region were "on it."

In good season in the morning, leaving Seven or Eight of the boys to cook beans and look after things generally in camp, we divided ourselves into parties of four and five and started in different directions toward the hills, on our Prospecting Tour.

Sam Batchelder, John Murphy, Wm. Cole, and myself went up the Beaver Head and crossed that stream to its North side, just below where Rattle Snake Creek empties into it. We sunk a hole near the bluff of rocks on that side and some distance from the River, and another in the gravel bottom through which the Rattle Snake runs, before reaching the Beaver Head River. We found some gold but not in quantities sufficient to warrant working. To the west, at a distance of three or four miles, the Beaver Head flowed out from the hills, and on its left bank, looking in that direction, a high rocky column arose, towards which we concluded to direct our footsteps.[47] I liked the appearance of the hills in that direction, which reminded me much of what I had seen in California, and Batchelder & Cole much regretted that we

---

47. The area identified by Purple, alongside the Beaverhead River four miles downstream from the mouth of Rattlesnake Creek, does not contain a single rocky column but rather the formation known as the Rattlesnake Cliffs—a name originally adopted by the Lewis and Clark Expedition. See Ingvard Henry Eide, *American Odyssey: The Journey of Lewis and Clark* (Chicago: Rand McNally & Company, 1969), 89.

Ed Purple and his companions prospected near Rattlesnake Cliffs (the formation right of center) located south of present-day Dillon, Montana. (Alfred E. Mathews, *Pencil Sketches of Montana,* 1886, plate 1)

could not stay for three or four days and prospect them, so well convinced were they that there was Gold Diggins in the neighbourhood.

I was however anxious to get over to Deer Lodge, and thought if my partner and self should follow our original intention of going to the Oregon Mines, which we had as yet by no means abandoned, we should need every day of pleasant travelling we now had to enable us to reach that destination before winter closed upon us. This was my reason for not being willing to spend more time than one day in Prospecting.

As we wended our way toward the Rocky column, we passed over a little flat on the River bottom covered with Cotton Wood trees, and Several varieties of wild currant bushes on which profusely hung black, red and yellow currants, now in season and perfectly ripe. So plenty were they that a person would have had little trouble to have gathered several bushels of this fruit in a very short time. They are slightly more tart than our common currant, but the berry was larger in size. On this flat we found several old deserted Indian Wakaups, or willow camps where the squaws had been dressing skins and I could not help admiring the taste of the Indians for selecting so charming a place as this for their Camping ground.

We lingered about in the shade picking and eating the delicious ripe currants till my friend Sam Batchelder accidently stumbled upon a Hornet's nest, the inmates of which settling upon his as-coxygis, broad and fair, being a man of 190 lbs. weight, sent him flying over the trunk of a fallen tree, his hands working like the arms of a windmill in his efforts to brush them off, and disappearing in the thick undergrowth that fringed the riverside with a celerity of movement equalled but hardly excelled by a frightened deer in a Canebrake. Taking the hint from Sam's short but decisive experience, we followed after at a respectful distance from the scene of his mishap, and joined him at the Rocky Column on the opposite bank of the River.

We found here a few colors of gold in the washings from the bed of the stream, but not enough for practical working purposes. Sam and I ascended the Rock, which rose perpendicular from near the Water's edge to a height of sixty or 70 feet, and from the top of which we had an extended view of the river and plain below, and our camp at a distance of four or 5 miles, and which we discovered had received an addition of a covered waggon since we left it. By this time it was late in the afternoon, and we started back to camp, upon reaching which we found Col. Sam McLean and Judge H. P. A. Smith there, who had come from Fort Limhi. The Colonel had a Span of small mules, and light waggon, which he afterwards used in prospecting the Country with, going after and follow-

ing men of horseback over hills and across gulches and ravines, when it seemed impossible for a waggon to escape
turning over. His great size, being a man of 270 to 300 lbs.
weight, prevented him from riding on horseback, and confined his locomotion to the waggon.

We found that the boys in extending their hospitalities to
McLean & Smith had let the Beans burn so badly, that we the
latest comers were unable to eat what was left at the bottom
of the Kettle, and so had to content ourselves with Bacon and
bread for our suppers.

Woods reported that he had found a fair prospect of gold
in the hills, in the southeast side of the camp, and exhibited a
small flake of Gold of three or four cents value, but thought
the spot where he found it was so far from water that it could
hardly be worked profitably by hauling the earth down to the
creek, which would have had to have been done to separate
the gold properly from it.

The others who had gone out had returned early to camp,
and making a sort of holiday of their resting spell, Tisdale had
tapped the whisky barrel, and Stokes had done the same thing
with a cask of Brandy, and the boys had had a good average
time of it, in which with our new friends, the Col. and Judge,
we participated till ten or 11 o'clock at night, and then sought
that refreshing slumber for which these mountains and valleys
are so justly famous.

August 1st. We rolled out at an early hour in the morning, Col. McLean & Judge Smith bidding us goodbye, pushed
on ahead for Deer Lodge. Upon reaching the crossing of
Beaver Head, a couple of miles from our yesterday's camp, we
met a half breed Mexican, Tom Lavaté [Lavatta], and two or
three half breed Indians with a small pack train, on their way
to Utah settlements for Flour.[48]

I was told by Frank that Lavaté was regarded as one of the
best guides in the Mountains, and subsequently learned that

his great experience and intimate knowledge of the Rocky Mountains north of us, and south as far as the Colorado River, justly entitled him to this character. He kindly offered to take letters for us to Salt Lake, which I hastily wrote; and placed in his hands for that purpose. Crossing the River, the water nearly reached the beds of our large Government waggons, and washed into the small ones without however doing much damage. The current is gentle and flows in an easterly direction. We reached the Big Hole or Wisdom River before sundown, having made about 18 miles today.

August 2nd. Moved up the Big Hole, which flows in a South of east course, and crossed it at the point where the next summer Fred Burr and Jim Minesinger erected their Bridge. This stream has a more rapid current and greater volume of water than the Beaver Head, and the bottom covered with boulder rocks (or large cobble stones) made the fording of it exciting and laborious. We put twelve and 14 yoke of cattle to the waggons, and getting on the downstream or lower side of the teams, to prevent their turning in that direction, towards which the current would nearly lift them from their feet and carry them, we succeeded in getting them all safely over. Some of our things were wet, and Horace Wheat's small waggon was nearly submerged by getting too far down the stream. We camped for the night just above the crossing, having made about 10 miles.

There was an abundance of drift wood on the bottom, and we built up a rousing fire to dry our blankets and bedding. On this flat were the remains of an old Indian Corral which had been used by them in Corralling their Poneys, and

---

48. Tom Lavatta, born in 1820, had guided early Roman Catholic missionaries in the northern Rockies and worked for the Hudson's Bay Company. Granville Stuart found him living on Cottonwood Creek in Deer Lodge valley, later the site of the town of Deer Lodge, in 1860. He died at Pocatello in 1896. (Deer Lodge) *New Northwest*, October 2, 1896; Stuart, *Forty Years*, 1:526.

Deer Lodge resident Granville Stuart drew this likeness of his neighbor Jim Minesinger in 1861. (Montana Historical Society Museum)

judging from its size they must have had a band of three or 4 hundred horses at the time of using it. Frank told me that near this place the Stuart Brothers, Bob Dempsey, and other Mountaineers camped in the winter of 1857-8.

In crossing one of the little runs just before reaching camp, one of my wagons, on raising from the opposite bank, sunk its lower hind wheel so deep into the mud that it tilted over and sprawled Sam Batchelder, who was riding on the load, onto the ground, and piling on top of him the Flour and other things from the wagon. There was a cross cut and whip saw, and several edged tools in this waggon, and for a moment we feared that Sam was seriously injured, as nothing but the hinder part of his person could be seen, as he quietly lay awaiting the removal of the weight that fastened him helplessly down. Throwing it off, we were glad to find that he had escaped with a few slight bruises.

Aug 3d. Keeping up the Big Hole a short distance, we turned up the Divide, or as Frank called it, Indian Creek, slightly altering our course to a more easterly direction. Our main course had been North during our entire journey.

Today we met a party of Seven men among whom was John White and a man that Frank called Raymond, and who he told me had lived for a year or more in the vicinity of Deer Lodge. These men informed us they were on a Prospecting tour to a small stream that flowed into the Beaver Head some fifteen or 20 miles above the crossing of the Road, and that they believed there was good Gold Diggins there, and urgently requested us to turn back and accompany them with our teams to the place. Says White, whom I afterwards knew, "You will soon be travelling back this same road, on your way to our Diggins."[49]

Had we possessed the least acquaintance with these men, we might have had sufficient Confidence in them to have gone along, but [since this request was] coming from Strangers we thought best to continue on to Deer Lodge. [The members of] this party were undoubtedly the first men that dug Gold on Willard's creek, or Grasshopper as it was called

---

49. John White and his associates, veterans of the Pike's Peak rush, actually made their gold discovery at Grasshopper Creek—where Bannack City soon sprang to life—on July 28, 1862, one week before Purple's date for this encounter. Regarding the White expedition, Bancroft, *Washington, Idaho, and Montana*, 621, carries a brief account, and later authors have not been able to expand substantially on his information. See for example Greever, *The Bonanza West*, 216-17. John White had come from Colorado with a party that included Charles Reville, William Still, Walter B. Dance, and one other. It was White, Reville, and Still who made the gold discovery, after the others had turned back. The identity of their guide, the man addressed as Raymond by Frank Truchot, is not clear. This account by Ed Purple is the only known contemporary source to declare that the party had along an experienced local guide. Biographical sketches of Dance and White appear in Thrapp, *Encyclopedia of Frontier Biography*, 1:373-74; 3:1552.

by them, and the discovery of which peopled this country with at least 15,000 inhabitants in the spring and summer following—and leading to the finding of rich gulches and gold placers, fixed permanently the status of Montana as a rich and valuable Gold & Silver producing Territory.

We made our camp for the night near the top of the Divide on the Creek, having travelled about 18 miles.

August 4th. After the Train moved out this morning, I started on ahead for Gold creek to make arrangements for its arrival there, and gather such information as I could of the country and prospects for Gold Mining.

I was soon over the Divide, and again on waters flowing into the Columbia. The Road was so level that it was difficult to establish the exact point from each side of which flowed the waters that mingled into two mighty Rivers, and emptied themselves finally into separate and distant Oceans. This is also the character of the Dry Creek Divide and many others that bound these successive valleys and Divide their waters.

After two or three hours riding I reached the First Crossing of Deer Lodge Creek, and in a couple more the 2d crossing. At the First crossing the course of the stream is westerly, but here and from the Head of the Valley it flows northward. Instead of going over it, I turned down the trail on the right of the creek, which runs along near the high plateau of land that came down near its banks on this side. On the left across the creek, and watered by many streams of pure clear water, lay the beautiful valley of the Deer Lodge. The luxuriant high grass which covered the bottom, shaded by the Cottonwoods that profusely lined the banks of its creeks and streams, afforded ample pasturage to the herds of sleek, round-bodied cattle that roamed at liberty over the plain; and I thought I had never visited a more charming place than this, not even in dear old and ever to be remembered California. As a grazing country, I think it unsurpassed by any I have ever seen.

Near one of the streams that flowed down from the distant mountain hills through the valley, I observed a mound of half egg shape form, that rose from the level plains around it to the height of 30 or 40 feet, and about the same distance across its base, and which I first supposed was the work of Indians. A subsequent visit discovered that it was the Hot Spring Mound, on top of which a warm spring of well shape rises nearly to the top surface, while surrounding it at the bottom are innumerable springs of water, of various degrees of temperature, the exhalations from which strongly resembles the smell of rotten eggs. The country for a short distance around the mound is swampy, and covered with a species of salt grass and tule, Somewhat of the character of the tule flag. The Mound is formed of a greyish red rock, specimens of which I brought away with me, and its lower portion is covered with patches of the salt grass which finds a sickly growth on its sides.[50]

I have been told that the Indian women frequently take their children to it when sick, and bathe them in its waters. The chemical properties of the spring have by some mysterious process of Nature formed the Mound, which is no doubt still gradually and imperceptibly to our observations receiving yearly additions to its height and size. The valley is said to have derived its name from its being a favorite resort for Deer in former times, but which are now seldom seen.

After dark I reached a log cabin the only occupants of which who made themselves visible were an Indian woman and two or three half breed Children. I enquired the way to John Grant's place, to which they replied in the Indian

---

50. This hot springs, marked on the 1859 Bureau of Topographical Engineers' "Map of the State of Oregon and Washington Territory" and reproduced in *Maps of Early Idaho*, today is the site of an upscale vacation facility, the Fairmont Hot Springs Resort, located off Interstate 90 between Deer Lodge and the Anaconda exit.

Ed Purple's "beautiful valley of the Deer Lodge" (Alfred E. Mathews, *Pencil Sketches of Montana*, 1886, plate 31)

tongue, and by making signs I understood it was a short distance down the small creek which I was then near by.

Spurring my poney along I soon reached a good sized log House, or rather two joined together, and reigning up asked permission to stay over night. To this Johnny Grant the proprietor readily assented, remarking "if I could put up with the fare he had to offer I was welcome." Sending out an Indian boy to unsaddle and turn in the grass my poney, he invited me to take a drink, which I find to be pure neutral spirits, but diluting it with water, quite palatable after my long ride of nearly 40 miles that day.

I then set down to supper, which consisted of bread and fresh meat, served up in Indian style, that is, all in a pile, to which I did ample justice. John Grant was born and has lived all his life, with the exception of occasional visits down the

River to Saint Louis, in the Mountains. His Father old John [Captain Richard] Grant, a Frenchman now dead, was an attache of the Hudson Bay Company, whose first wife was the Mother of John and one or two more boys, and several daughters. Old John's second wife was a half breed Indian, who had some advantages of education, and two or three of her daughters went to school I believe at Olympia, Washington Territory; and are quite respectably married. I heard while at Deer Lodge, and in the vicinity, that they were members of the Catholic Church.[51]

John Grant, or as he is called here Johnny Grant, is a man I should judge of 32 or 3 years of age, tall and well formed, with less of the Indian than American in his features, of a dark swarthy complexion, increased no doubt by exposure and his habits of life. He had I understood three Indian wives, and seven or Eight, perhaps more, children, that he treats with an affectionate fondness of manner that the children of many white men might envy. His little son David seemed to be the more especial pet of his father.

---

51. Purple errs concerning the elder Grant. Richard Grant was a Canadian-born Scot who served in the fur trade for the North West Company and the Hudson's Bay Company. In late 1841 he took responsibility for directing the Snake River trade from Fort Hall. His second wife, a native of the Pembina settlement on the Red River of the North and the widow of fur trader William Kittson, bore him three daughters and two sons, John and James. Following his retirement for reasons of health in 1851, he occupied the abandoned U.S. military post nearby, Cantonment Loring, and supported his family by trading cattle and horses with overland emigrants. In 1856, after the Hudson's Bay Company closed Fort Hall, he moved his winter base to the Beaverhead country. Late the following year Louis Maillet brought him news of the Utah War, coupled with an assertion that the Mormon church leaders planned to expand settlement into the Beaverhead region. Alarmed at the report, Grant moved his outfit to the Hellgate area with Maillet's help and established a ranch near modern Missoula. In June 1862, shortly before Purple's arrival at Deer Lodge, he died at Walla Walla while returning from a spring supply trip to The Dalles. See Merle Wells, "Richard Grant," *The Mountain Men and the Fur Trade of the Far West*, ed. LeRoy R. Hafen (10 vols., Glendale: Arthur H. Clark Co., 1965–1972), 9:165-86. See also Stuart, *Forty Years*, 1:125-26, 211; 2:97, et passim; and Thrapp, *Encyclopedia of Frontier Biography*, 2:580-81.

Johnny Grant, shown here in 1866, prospered in the Deer Lodge valley and extended a welcome to all who passed by. (Montana Historical Society Photograph Archives)

Here in this luxuriant grassy valley, abounding with game and fish—the finest brook trout I ever saw—possessed of large herds of cattle and Horses, surrounded by his half breeds, Indian servants, and their families, with a half dozen old French mountaineers and Trappers who have married Indian women for his neighbors, Grant lives in as happy and free a manner as did the ancient Patriarchs.

Let me wish, for the open hearted manner in which he bestowed his hospitality on the "stranger within his gates," that this pleasing life of his will not be invaded, and that he may long live in the enjoyment of the blessing with which a bountiful nature has so lavishly surrounded him.[52]

---

52. John Francis Grant, born in Canada in 1832, joined his father at Fort Hall in 1842, then struck out on his own during the winter of 1849–1850. He moved to the Beaverhead valley in 1855–1856, meanwhile participating in the cattle and horse trade with overland immigrants at Fort Hall. He wintered in the Deer Lodge area in 1857–1858, and settled there permanently in the fall of 1859. His ranch, located just north of modern Deer Lodge, became the center of a settlement sometimes called Grantsville or Grantsburg by Granville Stuart and others. Assisted by his Bannock and Shoshone wives and a large circle of relatives, he warmly welcomed all travelers, Indian and white. In 1866 Grant sold

August 6th. Stopping at Grant's place were five or six persons from St. Louis, who had just come up the Missouri River and over from Fort Benton.[53] From this circumstance and the attention with which I had been treated last night and this morning, I concluded his house was an Inn, and upon my leaving extended to him a $5 Gold coin, which he courteously but peremptorily refused to accept.[54]

Thanking him, I mounted my poney, crossed the creek and soon lost sight of his hospitable abode as I galloped down the Valley towards Gold Creek. Upon reaching the Little Black Foot, a small stream about 10 miles below Johnny Grant's, I found three or four log cabins, one of which, used as a store, contained a small stock of merchandise, Groceries, &c., owned by Captain Nick Wall, who had this summer come up from Saint Louis and established a Trading Post at this point.[55]

---

his ranch and cattle to Conrad Kohrs for $19,200 and returned to Canada. There he married Clotilde Bruneau, the daughter of a prominent judge. Grant died in Edmonton, Alberta, in 1907. With nine children born in Montana and seven more in Canada, he achieved a considerable posterity. Among other sources, see Bell, "Perry W. McAdow and Montana in 1861–1862," 48; Stuart, *Forty Years*, 1:126, 166-67; 2:97. For information on both Johnny Grant and Richard Grant, I am indebted to Lyndel Meikle of the Grant-Kohrs Ranch National Historic Site in Deer Lodge and to Will Bagley of Salt Lake City.

53. As early as June 29 James Stuart's journal notes the arrival at Deer Lodge of a small party from St. Louis who had come up the river by steamboat as far as Fort Benton. These men were traveling to the Salmon River mines, having never heard of the diggings at Gold Creek, but the two who tried the journey got hopelessly lost and returned to Gold Creek in late July. (One of them, Samuel T. Hauser, became a leading figure in the development of Montana's mining industry and political institutions.) By July 5, about forty-five newcomers were at work along Gold Creek. Stuart, *Forty Years*, 1:210-12, 214.

54. Conrad Kohrs, a German immigrant who became Montana's most prominent and successful cattleman, took over the Grant property in 1866. Thanks largely to Kohr's grandson, Conrad Kohrs Warren, this property is now the Grant-Kohrs Ranch National Historic Site. The two connected log cabins where Ed Purple enjoyed Johnny Grant's hospitality later became bunkhouses for Kohrs cowhands. Well preserved, these structures are situated close to the main house also built by Johnny Grant and subsequently enlarged by the Kohrs family.

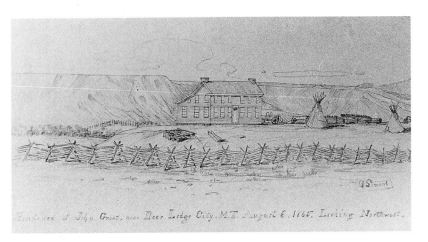

Ed Purple enjoyed Johnny Grant's hospitality in the two connected log cabins seen between the tepees in Granville Stuart's 1865 drawing. The big house was built in fall 1862. (Montana Historical Society Museum)

Crossing the Little Black Foot which comes in from an easterly direction and empties itself into Deer Lodge Creek, which is now called Hell Gate River, I continued my journey down the stream. Here the valley terminates and the River is hemmed in by the low hills and bluffs which reach on either side of its banks, and to follow down which with wagons involves the necessity of crossing three times before reaching Gold Creek, or American Fork, [at] which place I arrived before 11 o'clock in the morning, having travelled about 16 miles.[56]

---

55. Nick Wall, born in 1820, was a native of Alexandria, Virginia, who had earned his title as a steamboat captain, builder, and agent in the Mississippi River trade based at Galena, Illinois, a frontier lead-mining town that boomed in the 1840s and 1850s. He pursued his fortune as a merchant and miner in Montana's gold camps from 1862 until 1876, when he retired to St. Louis. Among his distinctions, Wall was one of the five founders of the Montana Vigilance Committee in late 1863. He died in St. Louis in October of 1880. Helena *Herald*, October 11, 1880; Thrapp, *Encyclopedia of Frontier Biography*, 3:1504-5.

56. The site of James and Granville Stuart's small settlement, previously known as American Fork, was renamed Gold Creek three weeks earlier, at the time of an election for a territorial representative and a slate of county officials as part of Missoula County, Washington Territory. Stuart, *Forty Years*, 1:213-14.

Captain John Mullan commanded the United States Army expedition that built the Mullan Road connecting Fort Walla Walla with Fort Benton between 1858 and 1862. (Montana Historical Society Photograph Archives)

Between Johnny Grant's place and [the] Little Blackfoot, I had crossed the Mullen Wagon road which was built by a Military Expedition under the Command of Capt. John Mullen in the fall of 1859, and which crosses Deer Lodge creek to the west side through the low bluffs, some distance from the River, down which it follows and crosses the American Fork at this point.[57] I found four or five Log cabins here, in which the Stuart Brothers, Fred Burr, Thomas Adams, Jim Minisinger, John Powell, Peter Martin a Mexican half breed, with their Indian families were living.[58] Finding Fred Burr, who I had become acquainted with at Fort Limhi, and ascer-

---

57. Captain John Mullan, a native of Virginia and a West Point graduate, arrived in the northern rockies as part of the Stevens northern railway survey expedition in 1853. During the winter of 1853–1854 he began the task of establishing a wagon road to connect Fort Benton, the head of navigation on the Missouri River, to navigable waters on the Columbia River. After a two-year interruption to campaign against the Seminole Indians in Florida, he resumed construction of the wagon road in 1858 and completed the task successfully in 1862. He resigned from the army and, after various failed enterprises, opened a law practice in San Francisco. He died in Washington, D.C., in 1878. Thrapp, *Encyclopedia of Frontier Biography*, 2:1032. For detailed description of the Mullan Wagon Road project, see Jackson, *Wagon Roads West*, 257-78.

taining that Jack Mendenhall, who was camping with some of his party close by, would soon be back from hunting up his Cattle, I unsaddled my poney and turned him out to feed upon the bunch grass which luxuriantly grew on the hillside and bottoms in the vicinity.

In a few moments Burr invited me to sit up to the table, and eat the food his Indian woman Fink had been busily preparing, without any intimation of whether I was hungry or not. And this is a custom which seems universally to prevail among the Mountain Men and their families, and speaks volumes for their kind heartedness and good feeling towards their fellows.

I was not hungry, as I had with me a day or two provisions, but accepted the proffered hospitality as an acknowledgement of my appreciation of the kindly feeling with which it was bestowed.

I remained all night with Jack Mendenhall and his companion, Bob Meneffee. Tom Pitt had started for the Florence

---

58. Minesinger, Powell, and Martin, like the Stuart brothers, Fred Burr, Thomas Adams, and other early Montanans previously identified, had made comfortable lives for themselves in the high country prior to the gold rush excitement. James Madison Minesinger, born in Tennessee in 1824, arrived in 1856 as a drover with Fred Burr's herd of cattle from Salt Lake City. With his Shoshone wife, he settled in Johnny Grant's little community in 1859. He mined briefly at Gold Creek in 1862, then moved over to the Big Hole valley, and in 1876 was living in Missoula. He later moved to Calgary, Alberta, where he died in 1894. John W. Powell, Minesinger's partner in the cattle business, was a native of Virginia, born in 1834, who also arrived in Montana with Fred Burr's cattle herd in 1856. Married to a Bannock woman and camped with her people near Fort Limhi in February of 1858, he warned the Mormon colonists of their danger some weeks before the Shoshone and Bannock attack, only to have his warning disregarded. Powell did some prospecting, mined for a time at Gold Creek, and remained in the Deer Lodge area for several years. He was shot and killed in a personal dispute in 1879. Less is known about Pedro Martinez, who anglicized his name in the mountains to become Peter Martin. At Fort Union in 1853 Isaac Stevens hired him as a hunter for the northern railway surveying expedition. He moved from Fort Union to Johnny Grant's neighborhood in 1859. All three men are mentioned frequently in Stuart, *Forty Years*. For biographical sketches see Thrapp, *Encyclopedia of Frontier Biography*, 2:948, 994; 3:1168-69.

James Stuart, pictured here on December 1, 1864, was Granville Stuart's brother and also lived in the Deer Lodge valley. (Montana Historical Society Photograph Archives)

Mines. H. F. Morrell was buying Cattle of Johnny Grant and butchering them for the Emigrants and a few men who were up the Creek working and prospecting for Gold. Bosdick, Purcell, and Haley with their partner had gone prospecting.

I learned that about 4 miles above, Granville Stuart, his brother James Stuart, P. W. McAdow, Sternie Blake, Fred Burr, and John Powell had been digging for Gold on a small creek called Pioneer Creek, which runs into the American Fork near its head, since sometime in April. They were using Sluices in washing the earth, and though not every successful in their Mining operations, have the honor of being the first white men engaged in practical Mining in this Territory.[59]

---

59. Granville and James Stuart record in detail the varied success of these operations. Stuart, *Forty Years,* 1:205-21. See also Bell, "Perry W. McAdow and Montana in 1861–1862," 41-53. For biographical sketches of Perry W. (Bud) McAdow and Abraham Sterne (Sternie) Blake see Thrapp, *Encyclopedia of Frontier Biography,* 1:125; 2:889.

Some of Ed Purple's contemporaries headed west over the Bitterroot Mountains to the Salmon River goldfield and the mining town of Florence, Idaho, shown here circa 1896. (Idaho State Historical Society, Boise)

An eccentric white man known as "gold Tom" had long prior to this, running back two or three, or more years, prospected the country laying in and between the Boulder and Prickly Pear Creek, and the Deer Lodge, but had accomplished little more—of which he ever gave any account—than the finding of small quantities of Gold, too small to engage attention in working.[60]

It was currently reported, and my special informant of the fact was Major Wm. Graham, one of the most intelligent Mountain Men I ever knew, that a half breed named Penetzee was probably the first person who discovered Gold—some time in 1851-2—in the region included within the present boundaries of Montana Territory. His explorations were limited and probably confined to the bare discovery of the precious metal.[61]

From all I could learn of what was actually being done on Pioneer Creek in the way of Gold Mining, I concluded to bring the Train here. Some Coloradians, among whom was Andrew Murray, had commenced work on a small creek south of Pioneer, a short distance from it, but with a reticence characteristic of miners generally. It was difficult to learn how well or poorly they were doing. They had named their diggins Pike Peak Gulch, an honoured name which it still retains.[62]

---

60. For the career of Henry Thomas, nicknamed Gold Tom by Granville Stuart, see Stuart, *Forty Years*, 1:161-65.

61. The fullest account of gold discovery by Francois Finlay, known as Benetsee, appears in the writings of Granville Stuart, excerpted by editor Paul C. Phillips in a lengthy footnote in Stuart, *Forty Years*, 1:137-40. Purple here reflects Stuart's views. See also Granville Stuart, "A Historical Sketch of Deer Lodge County, Valley and City," *Contributions*, 2:121-23; Paul C. Phillips and Harrison A. Trexler, "Notes on the Discovery of Gold in the Northwest," *Mississippi Valley Historical Review*, 4 (June 1917), 89-97; and William S. Greever, *Bonanza West*, 215-16.

62. Granville Stuart described the start of these operations in his 1865 publication, *Montana As It Is*, a description reprinted in Stuart, *Forty Years*, 1:212-13.

August 6th. I started back this morning to the Train, which I met in the forenoon coming down nearly opposite Johnny Grant's, and which moved down and camped for the night on the Little Black Foot, near Capt. Wall's store. A mania had seized upon some of the persons newly arrived here to build a town, and the bottom had been surveyed off for a mile or more into streets, squares, and building lots. This scheme was being Engineered by members of the American Exploring and Mining Company, recently arrived from Saint Louis via [the] Missouri River, one of the principals of which, Mr. Henry C. Lynch, tried to induce us to remain, descanting on the beauty of the place, its natural advantages for wood and water, its proximity to the Pike Peak's Gulch Diggins, and offering us several lots if we would stop and build a house on one of them. The plan of the new town was elegantly mapped off, on paper, but as I had lived many years on the waters that laved the Tule lands round about New York on the Pacific at the head of the bay of San Francisco, and with some reason knew something about founding new cities in a wilderness, I declined the offer.[63]

P. C. Woods, Who was acquainted with Lynch, fell in with his views and remained, as did also Gamble & his son, Stokes, Rhinehart, Dewes, Harry Gray, John Knowles, John Tripp, Sr. & Junior. The last two a couple weeks after left and went to Fort Walla Walla in Oregon.

The balance of the Company joined me and went down to Gold Creek, which we reached—after tipping over one of my heavy wagons—late in the afternoon of the 7th August, just two months from the time of our departure from Salt Lake City.

---

63. As Purple relates, the American Exploring and Mining Company, an organization founded in St. Louis, attempted briefly a townsite promotion at the present site of Garrison. On formation of this firm, see the autobiographical account by Judge Francis M. Thompson, "Reminiscences of Four-Score Years," *Massachusetts Magazine*, 5 (1912), 138, 161.

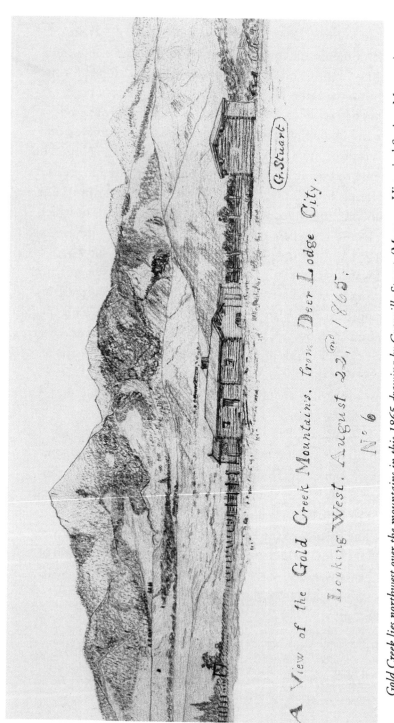

Gold Creek lies northwest over the mountains in this 1865 drawing by Granville Stuart. (Montana Historical Society Museum)

CHAPTER FOUR

⚜

Gold Creek
August 7 to December 15, 1862

By the time we reached Gold Creek, Tisdale and I had about made up our minds not to go any farther north this season, and for the present all regarded this as the end of our journey. The company in which we had travelled was broken up, and the small measure of restraint imposed on each individual for the preservation of order, harmony, and safety of the Train while on the Road was now thrown off, and each one left free to secure in his own way the best means of obtaining a livelihood and support.[64]

The next day Dr. Ford, who had driven one of my teams up to the time of the breaking down of the wagon on Dry Creek, attached himself to a passing Train, and left for Fort Walla Walla. He was a regular educated Physician, and it was his intention either to practice his profession in Oregon, or teach school there, in which latter occupation he had been engaged, though not a Mormon, in one of the small settlements in Salt Lake Valley previous to starting with us on our

---

64. On August 8, the day after Purple says his wagon train arrived at Gold Creek but probably a week following the actual date, James Stuart made the journal notation that he and his partner, Frank Woody, had bought "twenty three gallons of Valley Tan [whiskey] at six and a half dollars per gallon from Purple & Co." James and Granville Stuart Diaries, Beineke Library, Yale University, New Haven, Connecticut. Editor Paul C. Phillips omitted this entry in his published version of the Stuart materials.

present trip. He was a mild, pleasant man, of medium size, slow and formal in his manners, without much energy, and addicted to indulging in his Cups slyly—a bad sign—but which had specially endeared him to the only Irishman in our party, John Murphy.

[At one point] We had been several hours in ascending Cady hill, on our journey, doubling and trebling the teams to each wagon, filling in ruts made by the rain, and cutting down the side hill in places for the better passage of our wagons. Joseph Vassar, one of the party, a Frenchman [and] a Blacksmith by trade, low set in person but stout, and active as a cat, a well intentioned fellow, prompt in doing whatever he saw needed to be done, but withal extremely officious in his manners—he had become irritated at the slow movements of Dr. Ford in the work on Cady Hill, and had let fly at him a volley of "Sacre bleu de expressiones" which astonished the Doctor, and raised to a fighting pitch the Milesian wrath of his friend Murphy. Arriving at the scene of the controversy, I sharply rebuked the Frenchman for his meddlesomeness, and reminded Murphy of the conditions upon which he was a member of the party, made before leaving Salt Lake City; this was that if he became a party in any quarrel, or fight, on the journey, I would expel him from the Train. Upon this the matter was dropped, and good feeling apparently restored to all parties, which lasted to the end of the journey.

On the day Dr. Ford left us we partly unloaded our wagons and placed our things under a tent made from Extra Wagon covers, which we erected near the Creek and a short distance from an Indian lodge, belonging to Burr, which Jack Mendenhall & his partners were using. Wheat & Vassar were camping with Tisdale, Batchelder, Colo. Murphy and myself, which now comprised our entire party. About sun down, as I was seated at the mouth of Jack's Lodge, I heard loud voices and an angry discussion at my Tent to which I hurried and

found my Partner and Murphy engaged in a word controversy, of which I appeared to be the subject. I determined to
take the initiative, and hitching up my pistol, always at my
side, that great leveler in western life of all muscular distinctions, I told Murphy who I saw had been drinking, to begone
from the place.

Retreating to the opening of the tent, he blarney'd out his
rage, and contempt for favorites, and flourishing a hickory
pick handle over his head—a nice 'Shillah Schtick'—he swore
"he'd bate the head off the thaiving Frog ating frenchman, so
he would. The dhirty Spalpeen, putting on his French airs
wid him. Oh be gorra I'm the By that 'ill show him. Where is
he? Be Jasus I'll not lave a whole bone in his body—as sure as
his Mutter borned him. And may the divil fly away wid his
Sowl."

I told him he could have no fight about my Tent—and he
knew me well enough to believe it. For cursing as only a
drunken Irishman can curse, he started off up the Creek, to
Stuart's Chebang (whiskey shop). Where Jo was during this
time I did not know, but in an hour or two Murphy found
him back of Mendenhall's Lodge, and beat his head and arms
black, till hearing the fracas, Jack, Burr & myself first arriving
at the scene of the disturbance, pulled him off and wrenched
the Pick handle from his hands, in doing which as the other
object of his drunken rage, he threatened me with direct
vengeance.

Going to him in the morning and telling him that I did
not propose to live in dread and fear of his, or any body's else
threats, I expelled him from our mess, and offered him direct
instant personal satisfaction for any grievance he had towards
me. I done this quietly and alone, and am satisfied I saved
myself in so doing much trouble.

He acknowledged his error, swore he would never drink
another drop of whiskey, which he kept till after we went to

Bannack, blarneyed the Frenchman (who had really acted cowardly) with expressions of sorrow for the aches and bruises he had bestowed on him, and peace "and order once more reigned in Warsaw." I never had the least bit of trouble with him afterwards.

We found at Gold Creek upon our arrival there, in addition to those already named, the following persons: H. M. Mandeville, Geo. Colburn, Maj. Wm. Graham, Ned Williamson, Wm. Hamilton who called himself Wild Bill of the Mountains when drunk, Thomas an Irish boy with Mendenhall & Minnifee, William Carr a blacksmith, Frank Woody, Bar Keeper for Stuart Bros., Al Clarke, Dr. McKellops, a Saint Louis Dentist & Rebel, Geo. Parker, Frank Louden, & a few others.[65] The last four had come up from Saint Louis this season, and had the finest outfit for travelling I ever saw in the mountains. McKellops, Parker, and Frank Louden soon left for Fort Walla Walla, on their way via San Francisco for St. Louis. Al Clarke remained, and was employed as clerk in Wall's store at Little Black Foot.

[At Gold Creek] People were continually arriving in small numbers, some stopping a few days to prospect the new mines, and scattering themselves over the hills and mountains, others to whom Gold Digging had no charms, or who were doubtful of its being found in sufficient paying quanti-

---

65. Purple here again demonstrates his propensity for recording names. On Graham, Williamson, Hamilton, and Woody see biographical entries in Thrapp, *Encyclopedia of Frontier Biography*, 2:578, 610-11; 3:1575-76, 1595. See also W. T. Hamilton, *My Sixty Years on the Plains, Trapping, Trading, and Indian Fighting*, ed. Donald J. Berthrong (1905; reprint, Norman: University of Oklahoma Press, 1960), vii-xvi. (Unfortunately for present purposes, Hamilton sums up his experiences during Montana's gold rush era in less than two sentences.) For a biographical sketch of Albert G. Clarke, see *Progressive Men of the State of Montana* (Chicago: A. W. Bowen & Co., [1902]), 152-53; and Bancroft, *Washington, Idaho, and Montana*, 765. For an obituary of George Parker, see Helena *Herald*, September 10, 1886.

Gold Creek barkeeper Frank Woody, pictured here circa 1890, served as Missoula County clerk and recorder from 1866 to 1880 and in 1892 was elected judge for the Fourth Judicial District. (Joaquin Miller, *Illustrated History of the State of Montana*, 1894, p. 708)

William T. "Wild Bill" Hamilton is pictured in his later years in his book. (Sixty Years on the Plains, 1905, frontispiece)

Gold Creek store clerk Al Clarke, shown here about 1890, opened a store of his own in Virginia City in 1864. He later became a prominent Helena merchant. (Progressive Men of the State of Montana, [1902], p. 153)

ties, continued on down Mullen's Military road, to the more inviting settlements of Oregon.

As yet the capabilities of the country for agricultural purposes had not been thoroughly tested, and except the sowing of a few oats by Grant and Rob't Dempsey, who lived a short distance down the Hell Gate, and the planting of a few potatoes by some others of the old Mountaineers, no cultivation of the soil had been attempted. The season was considered too short for raising corn and other cereals and garden vegetables except those of the most hardy varieties, and to which therefore little attention had been paid.

The nomadic life of the old residents, with their herds of Horses and Cattle nearly supplying all their wants for food, and as a means of Barter with the upper Settlements of Utah and with passing emigrants, furnishing them ample supplies of Breadstuffs, assimilated them to the Indian in their habits and tastes, and they found neither necessity or inclination in tilling the rich soil of the valleys by which they were surrounded.

The creeks, rivers, and mountain lakes were filled with the choicest brook trout, food which the gourmand might envy; while the fur of the mink, Beaver, and otter, and the peltry and skins of the Antelope and Deer, which roamed in great number over the hills and valleys around them—dressed and elaborately adorned with beads and fancy colored cloths by their Indian women—afforded them all the clothing they required, even in the rigorous winters of this northern latitude.

We now set to work prospecting the country round about the head of Gold Creek, above Pioneer Gulch, where in a few days, on the 10th August, we found Gold on a little run which emptied into Pioneer [Gulch] well up in the hills, which we named California Gulch. The Gold was coarse and heavy, and from two or three pans of earth first washed we obtained nearly a dollar, which I still preserve as a remembrance of the time and place it was found.

Willingly imparting the information of our discovery, for there was [an] abundance of territory after our own claims were located, we sent word to our friends who had stopped at Little Black Foot; and Dewers, Rhinehart, Harry and a number of others soon came over and commenced work on the Gulch.

Soon after, a report reached Gold Creek that Rich Diggins had been discovered on the east side of Deer Lodge Valley, on the Prickly Pear and Boulder Creeks. Such reports as [these] fly from mouth to mouth over a large section of country. [They] lose nothing but are rather magnified by the distance they travel, and in the unsettled condition in which nearly everybody was then in, was willingly believed, without much inquiry into its authenticity or correctness.[66]

John Powell, one of the old residents of the valley and an unprincipled fellow, who [had] done much to fan the excitement among the Pilgrims (a name by which the old Mountain Men designated men who were on the plains for the first time) by circulating stories of the extent and richness of these reported mines, told me some weeks after that he had done it to get the country prospected. Our boys took the fever and started off with a crowd from different parts of the valley above for the New Mines. I remained at Gold Creek. In five or six days they returned, not having been able to find better diggins than those they left on California Gulch. The place they visited was called by the Stampeders Humbug Creek, a small tributary of Boulder Gulch.

A few persons remained in the Prickly Pear Country and along Boulder Creek, and received additions to their numbers

---

66. Stuart's journal chronicles on August 7 the arrival of this news, which started "a general stampede over the mountains." Stuart, *Forty Years*, 1:216-17.

from Captain Fisk's Train, under the auspices of the U.S. Government, who reached that point in September from Saint Paul, Minnesota. There were over a hundred persons in his train, among whom were several families who settled themselves permanently in the country and are among the most respectable in the Territory.[67] Two months later the Prickly Pear District was deserted for the rich placers of Willard Creek, at Bannack City.

Another report reaching us that Discoveries of Gold had been made at the head of waters flowing into Bighole, North of Fort Limhi, and wishing to obtain as much information as possible by personal inspection before settling fairly down into winter quarters, I resolved to go over.

On the 30th August, accompanied by Fred Burr as guide, Tom Pitt, P. C. Woods, H. M. Mandeville, Geo. Colburn, [and] Granville Stuart, we left Gold Creek for that locality.[68] Burr provided two extra Horses to pack our Blankets, and provisions necessary for the trip. These were packed in parfleches of skins prepared from Raw Hide, from which the hair is removed and cut in form to fold compactly, by the Indians and Mountain Men, and are impervious to rain except it should be severe and continued for a number of days, when they become flabby from the absorption of the water. As a receptacle for camping utensils and provisions, these parfleches hanging from the cross trees of the pack saddle, on each side, and bound by a Rope passing around the animal's body to prevent shifting, are much more convenient (being more portable) than the basket-shaped Spanish pan-

---

67. On the Fisk expedition of 1862, see Helen McCann White, ed., *Ho! For the Gold Fields: Northern Overland Wagon Trains of the 1860s* (St. Paul: Minnesota Historical Society, 1966), 23-72.

68. Frank Woody had brought the news from Hellgate on August 19. In his journal entry, James Stuart gives August 22 as the date of departure for this party, with August 30 as the return date. Stuart, *Forty Years*, 1:217-18, 221.

Granville Stuart, pictured here circa 1870, ranched in the Deer Lodge valley and chronicled Montana history in his books and drawings. (Montana Historical Society Photograph Archives)

niers made of the same Material, and used for like purposes by the Mexicans of California, Sonora and New Mexico.

We took a south course and travelled along over the high plateau of lands that reaches from the base of the Mountains to the Hell Gate River, crossing Rock and Dempsey's Creeks and other small streams near the points where they flowed out from the western ridge of Mountains. These ridges were covered with Mountain pine, the streams lined with willow and some cottonwoods, and the small valleys covered with high grass. We saw large flocks of Antelope scampering over the country and disappearing in the distance like the shadow of a Sun cloud, down the ravines of the table lands. Before we reached Dempsey's Creek we left the plateau, and were in the bottom land of the magnificent Deer Lodge Valley.

We camped in the afternoon near an old cabin on Race Track creek, at the base of the Ridge, and amused ourselves for an hour in catching the speckled trout with which the waters of this valley are filled. This is indeed the paradise of a

Fisherman. It must not be supposed from their abundance, that the sport of catching is diminished, for the Trout here still retains his gamy characteristics, and the finest specimens are only caught by Skillful angling. Securing as many as we wanted of these fine fish, from a half to 3 lbs. weight, for Supper and breakfast, and some to take over the Mountain, we ate our supper, smoked our pipes, and told stories around our campfire, for the night air of the Mountain was even at this season keen and sharp—and not the Monarch on his throne was happier than we.

By sun rise we were off, travelling nearly west, up a small fork of the stream on which we camped the night before. The heavy dew hung in the tall grass and glistened in the sun shine and from the bushes along our path; the charm and beauty of which we could have well dispensed with, for it made us "sensible to feeling as to sight," as it showered down and wet our feet encased in moccasins, which being light and easy we preferred to boots for riding. We found large quantities of ripe choke cherries as we rode along, of the same character and appearance as those found in the Atlantic States.

We soon reached the thickly wooded portion of the Pass, and followed along an old Nez Perce Indian Trail which evidently had been made years before.[69] The deep silence of the forest was only disturbed by the tramping of our horses' feet, or the occasional flight of a specie of Grouse or Pheasant, which would sit unobserved on the limbs of a tree until we

---

69. Leaving the Deer Lodge Valley, drained by the Clark's Fork of the Columbia, the party headed up the Mill Creek drainage, along the route taken today by state route 274, which leads southwest from Montana Highway 1 across the divide to intersect with Montana Highway 43, the road that now accompanies the Big Hole (Wisdom) River as far upstream as the town of Wisdom. Due to later clear-cutting associated with the Anaconda corporation's smelting operations, the mountainsides in this locality no longer have the timber cover Purple described but a thick growth of willow, alder, and birch along the streams makes horse travel difficult.

were close to it; and then rise so suddenly on its wings, and with so much noise, as to make us think the limb of a tree was falling on us. I found it impossible to keep from starting up whenever these little birds took wing.

For three or four miles our path lay over logs and trees which seemed to have been blowed down by the wind a long time ago. They were so thickly strewed upon the ground and crossed and piled over each other that I first thought it utterly impossible to find passage over this immense and heaped up mass of fallen timber. There was no avoiding or going around it, and Burr took the lead, disdaining to [go on] foot an inch of the way, and the entire party followed suit. My Poney was as surefooted as any Horse that ever pressed a hoof to the ground, but it needed that and all his nimbleness to pass with safety this wooden barrier. For nearly two hours we floundered over this forest windfall before reaching a clear path. One of the Pack horses was badly lamed, and Colburn received some bruises from falling from his horse, the result of a misstep.

We were now soon over the pass and on the east side of the Mountain Ridge, going towards the Big Hole River, before reaching which we camped for the night on a small stream flowing into it.

The next morning we found the country round us covered with a white frost, but which rapidly disappeared as the sun rose, and before noon the atmosphere was warm and pleasant. In the fore part of the day, as we were jogging along in Indian file, Burr ahead, I observed him lay at length on his Horse's back and motioning with his hand for us to stop, he slid from his animal and crouching low to the ground informed us that just ahead was a Grizzly Bear digging Camus root. Dismounting, three of us with our rifles crept up cautiously toward the Bear, who was hidden from view by an elevated point. Being to the windward of us, he could not scent us, and probably having no idea of being disturbed in his morning meal, we approached to within fifty or sixty yards of where he was, when he turned, raised himself in a sitting posture on his hind feet, at which instant we leveled our pieces and fired. Knowing they were difficult to kill, we expected that if we only wounded him, he would charge upon us, in which condition they are extremely ferocious and dangerous; [and so] we turned and hastily mounted our horses. The Bear gave a grunt similar in sound to that of a frightened Hog, and rushed off in the opposite direction, the blood from his body and broken fore leg trailing the ground as he made his way towards the timber at the upper end of the valley, where we had neither time nor inclination to follow him.

This was the only wild Bear I ever saw in the Rocky Mountains, [since they] are not numerous in this locality. In the winter of 1857-8 I met one—a large Grizzly—four miles below the Calaveras Big Tree Grove in California, who hurried out of sight as soon as he saw me. I am of the opinion that the age of this animal has something to do with its proverbial ferocity, or that there is a difference in this respect in individual specimens, as there is in the dispositions or tempers of men.

Soon after our adventure with the Bear, we reached the River, which flows south of east; the valley of the Big Hole is circular in form and thirty-five or forty miles across, but unlike Deer Lodge in its features and general character. The land is principally table land, sloping down from the high ridges of mountains that surround it, and from which there are numbers of small streams rushing through to the River. There are numerous small grassy valleys on these streams, but being of greater altitude than the Deer Lodge, and many degrees of temperature colder, are valueless, save for grazing purposes.

On making our noon camp, we got our fishing lines out but met with so little success in our attempted raid upon the finny tribe [that] we soon abandoned our efforts. It is a notable fact that while the waters that empty into the Columbia abound with fish, particularly brook trout, these are seldom caught on the tributaries of the Missouri.

We camped for the night ten or twelve miles from the River, on a small creek and near a Beaver Dam which these industrious and Skillful animals had thrown across it. During our travels we passed many of these, which in this valley are chiefly composed of willow twigs, cut off by the Beaver's teeth, and carried in immense quantities to the point where they are constructing their works. They are laid and interwoven thickly together, and spread over with dirt and mud, for which last operation the Beaver is supplied with a long trowel-shaped tail, which he uses for the purpose. Upon streams lined with large trees may be frequently seen stumps 2 & 3 feet in diameter which have been gnawed off by these Creatures, the bodies and limbs of which have been used by them for Dam Building. Their dams are as strong and tight as they could possibly be made by human hands, and the spring freshet pours its waters over their tops without material damage.

The next day, Sept. 2d, as we left the valley and entered the foothills of the western mountain ridge, we passed an Indian Corral, which must have enclosed 8 or 10 acres of ground. It had been built many years ago, of Pine poles which the adjacent forest supplied, and for strength and size and the large amount of labour and time consumed in its construction, far exceeded any work by Indian hands I ever saw in this Country. Burr informed me that notwithstanding its strength and the precautions taken by the Nez Perce—who built it—in guarding their stock, a party of Snake & Bannack Indians stole them out one night, and though closely pursued crossed

the Snake River with the entire Band of Horses the next day, a
distance of 80 to 90 miles, and safely escaped into their own
Country. About 4 o'clock we encamped on the Gulch, near
where Gold had recently been discovered, and to examine
which was the object of our journey.[70]

Sept. 3d. We spent the day in looking about the mines.
The principal diggins are situated on a small run of water,
between two small ridges or hills near the top of the mountain, which at this point is flat and in places quite marshy at
this season of the year. The country is thickly covered with
Pine trees whose drooping appearance indicates the presence
in winter of heavy snows which cover them, on this mountain
height.

There were some twenty-five or thirty men here, who had
made arrangements for wintering in this wild and secluded
place. They had come over from Fort Limhi, and were principally from Colorado. A few, perhaps half a dozen, had found
their way here from the Oregon Mines, and were from the
"other side," a term in Common use to designate the place
where Oregonians and Californians came from. From those at
work I gathered the information that they were making three
& four dollars a day, or a quarter of an ounce of Gold to each
man.

The country for two or three miles around had been
pretty thoroughly prospected, but the paying diggins were
limited, and confined to the run I have described. There was

---

70. Purple and his companions had apparently reached the Pioneer Gulch
diggings, located on Pioneer Creek and Nugget Creek, and reached by the
Gibbonsville Road (Forest Route 624) approximately eighteen miles west-
southwest of Wisdom, Montana. Although surrounded by federal forest lands in
the Beaverhead National Forest, the historic mining area is privately patented.
For a brief account of early mining activities in this region, see Oren Sassman,
"Metal Mining in Historic Beaverhead" (master's thesis, Montana State University, 1941), 42-46.

the appearance of Coal in a number of places, and in one locality on the right of the trail we came up, we saw some specimens thrown from the prospect hole of a miner.

The conclusions we arrived at were that at present these mines could not be worked profitably, owing to the difficulty of approach with waggons, and their distance from any point that could become, for some time at least, a central depot for provisions, clothing, tools, &c., necessary for the support of a Mining population.

Another and greater objection was the shortness of the working season, which from the altitude of their location could only extend to four or five months during the entire year. Even at this season water congealed at night, and I did not doubt that within a month the cold would suspend mining operations entirely.

Late in the afternoon we started on our return to Gold Creek, proceeding as far as the old Nez Perce corral, where we camped for the night.

Sept. 4th. The travel today was without special incident, save that shortly after making our night camp, before sundown, in the Big Hole bottom we were visited by one of a party of three Frenchmen who had camped a short distance below us. He was an old man upwards of 60 years old, and apparently indisposed to garrulity, but informed us, after introducing himself as an old mountaineer, that he had trapped Beaver on this bottom 40 year before. He and his companions were travelling about seemingly from habit and the love of adventure, which the old man's age had not yet sated or rendered stale. He took little or no interest in the accounts of Gold we gave him, and was apparently indifferent to everything but his Ponies, Which he watched with careful eyes, feeding around his camp. Upon leaving he told me he thought he should winter in Deer Lodge at Johnny Grant's, with whom he was acquainted. I never met him after.

Sept. 5th. We camped for the night well up on the ridge that divided us from Deer Lodge, having spent some time in prospecting on the little stream where we located our Camp.

Sept. 6th. We crossed over the Divide and across the fallen timber, as before without accident, and reached Race Track Creek about 4 o'clock pm. From this point our entire party left Burr & myself, and went down to Johnny Grant's and [Little] Black Foot. We took the trail we came up on near the foothills, for Gold Creek.

Riding along we chatted of former travels and the causes which led us first to visit these Mountains. Burr was an old resident, had an Indian family consisting of a Snake Indian woman he called Fink, a little girl of 4 or 5 years of age, Virginia or Jinny so nicknamed, and a boy born a few days prior to my arrival at Gold Creek, and who was named Dick. He was a son of Gen'l. Burr, in 1855 the U.S. Surveyor General of Utah Territory. His father left Salt Lake in 1857, owing to the trouble Brigham Young was then having with the U.S. Government, and to correct which President Buchanan sent out a portion of the Army to that place.[71] Young Burr instead of accompanying his Father to Washington City where his family resided, with a few friends (among whom was John Powell) started north towards the Oregon settlements, reached Deer Lodge Valley, and having a stock of cattle & horses and abundance of provisions, wintered there.

---

71. At this point Purple inserts the following footnote: "Upon hearing that the U.S. Army was on its way to Utah, Brigham drove all who were not members of the Church out of the Territory. A few private individuals remained but were kept under surveillance. Of these Jack Mendenhall was one." On the Utah War see Norman F. Furniss, *The Mormon Conflict, 1850–1859* (New Haven: Yale University Press, 1960). As Furniss recounts, Surveyor General David H. Burr arrived in Salt Lake City in September 1855 and almost immediately alarmed the Latter-day Saints by reporting to his superiors that the Church had illegally appropriated public domain. Burr became, in Furniss's phrase, a stormy petrel in Utah, using his authority to harass the Mormon leadership and calling for direct federal intervention to coerce the Saints.

Being pleased with their mountain home, their love of hunting gratified by the vast quantity of game around them, they took Indian women for wives and each succeeding year lessened the desire for civilized life.

The rapid influx of population in these sequestered wilds was daily changing their habit of thought, and reviving with such men as Burr the recollection of the home circle, advantages of education, and early influences of refined Christian civilization, to secure which for his children he was most anxious. "I like my children," says he, "as well as if their Mother had been the finest lady in the land," and he felt the paternal obligations toward them as binding as if no copper coloured face in playful glee turned up to his its childish countenance—half his own, yet unmistakeable in its Indian character—with his own blood coursing through its veins, and bone of his bone, "flesh of his flesh."

"I am in the Boat, though, and must do the best I can," continued he, "I will send Jinny to St. Louis and have her brought up in some good family—it would never do to take her and Dick to Washington. The old woman would raise the devil to be separated from 'em, but she'll have to stand it somehow." But no such test of her affection was ever made, and the family still lives unbroken in Montana.

How many fathers proud of their sons, the expected solace and staff perhaps of their declining years, how many loving Mothers have waited long years for the return of the absent boy, who is kept back from their doting hearts by such an alliance as Burr has formed?

It seems to be the habit of Mountaineers in selecting Indian wives, either to take one who has been made a Captive, or one of another tribe at a distance from the one in or near which they reside. I suppose this is done to rid themselves of the woman's family relations, who would be certain at times to live off their relative. There is as much Cousining among the Snake & Bannack Indians as we see usually with the Irish, and is common to all Indian tribes I have ever known.

Late in the evening [of] Sept. 6th we reached Gold Creek. The people during our absence had been somewhat excited over the hanging of a youth, a mere boy of 18, and the killing of his companion, who were desperadoes and Horse Thieves from Elk City, Oregon. They had been pursued from Elk City by one Fox and another, from whom after receiving hospitable entertainment they had stolen some Gold Dust and the Horses with which they had fled the Country. The boy's comrade had been instantly killed by Fox while dealing Monte, a game of cards he opened the first night of his arrival at John Powell's log cabin. The boy was quickly tried by a Jury, and his reckless bearing, added to the evidence of his guilt as a Horse Thief, sealed his doom. He was hung from the poles and cross tree, set up to hang butchered cattle on—a short distance from the creek. He was bold and defiant to the last, asked no mercy, which some of the boys told me might have been granted had he not been so abandoned in feeling and vindictive in his threatenings. He refused to give his name, but said Virginia was the state of his nativity. This was the first hanging by a [jury] done in Montana.[72]

We now set about building a cabin and fixing up our winter quarters. In this work Jack Mendenhall and Bob Menneffer joined us. We erected a Log cabin about 38 ft. in

---

72. James Stuart, an eyewitness who recorded this episode in detail, provides a different description. He states that the guilty man's name was C. W. Spillman and describes him as "a rather quiet reserved pleasant young man, of about twenty-five years." After being sentenced to hang by a miners' jury, according to Stuart, Spillman showed the calmness of a brave man, with no reckless bravado or protests, "and walked to his death with a step as firm and countenance as unchanged as if he had been the nearest spectator instead of the principal actor in the tragedy." Stuart, *Forty Years*, 1:218-20.

John Powell, shown here circa 1875, mined at Gold Creek in 1862, then moved to the Big Hole valley and, later, to Missoula. (Montana Historical Society Photograph Archives)

length [and] 16 ft. wide, with an opening in the front of 10 feet, the whole covered with dry grass upon which was spread a layer of mud four inches thick. This roof made our cabin warm and dry, and being well chinked between the logs and plastered with mud, added to its warmth and comfort.

Jack & Bob lived in one end of the house, Tisdale, Sam & myself occupied the other. Our two fireplaces in each apartment were made of Adobes or thick mud, moulded in boxes prepared for the purpose, each one of [them] the size of four ordinary red bricks, and dried in the sun three or four days before using. These Adobes are extensively used for building by the Mexicans, and adopted by the Mormons in the construction of their dwelling houses at Salt Lake City and elsewhere in Utah.

Leaving this work nearly done for Menneffer and myself to finish, the boys resumed their Mining operations on California Gulch. Cole & Murphy built a cabin there, and Tisdale & Sam Batchelder done the same thing, as the distance (some 6 miles) was too great to travel from our Cabin on Gold Creek to and from their work.

Not knowing how severe the winter might be, and unwilling to lose our Cattle or let them suffer for food, I busied myself in building a corral and cutting grass on the Hell Gate River bottom below the mouth of Gold Creek. I had brought with me two or three Scythe blades and Snaths, and a couple of Rakes for Hay making purposes.

I hired Wm. Fairweather, a native of Nova Scotia, to assist me in this work, paying him $2.50 per day and board. He was a stout, rather thick set and well proportioned man, slow and deliberate in speech, a good worker, and became famous the following spring by discovering Gold on Alder Creek, where Virginia City now stands.[73] He worked for me about ten days, by which time I had my quarters well fixed up for winter.

The experience of hundreds of other men I have known in California, repeated in the history of this man, is suggestive of the frowns and favors that overshadow and lighten up the fortunes of a Miner. He was a common labourer now, working hard for small daily wages. In less than a year he had at his command so much or more ready money than any man in the Territory. But to follow his fortunes further would be anticipating a future leaf of this journal. Suffice it to say that the kindly feeling that sprung up between us was not altered when he became for the Pioneer, and for a brief time, the petted wealthy Miner of Alder Gulch.

It was now about the middle of October. The reports from the Rich Diggins of Grasshopper, which had been reaching us since late in August, were fully confirmed. Tisdale, Batchelder & the other boys, after building their cabins and taking out $2.00 or more [a day] from California Gulch, had determined to go over to Bannack. In the meantime Jack Mendenhall, Pitt, Morrell, and Menniffer had concluded to leave for Salt Lake City, and nearly all the old

---

73. For a biographical sketch of Fairweather see Thrapp, *Encyclopedia of Frontier Biography*, 1:478-79. See also an obituary in (Virginia City) *The Montanian*, September 2, 1875.

residents of Gold Creek had decamped for the new Gold mines. Jim Stuart, Worden, Reese Anderson, Fred Burr & myself were about the only ones left. Pike Peak's Gulch was deserted.

A few straggling miners from the Oregon Mines passed every few days through the place on their way to Bannack. The quiet of the neighbourhood was occasionally disturbed by small bands of the Flat Head Indians returning from their fall Buffalo hunt in the Yellowstone Country. These Indians belonged to the Mission and church called St. Mary's, established down the Hell Gate River 75 miles from Gold Creek by Father DeSmet, a pious and zealous missionary of the Catholic Church.[74]

During one of their visits I had pointed out to me an elderly man, whose light eyes and complexion proclaimed the white blood in his veins, and who Burr informed me was the reputed son of Cap't. Clarke of Lewis and Clark's famous Expedition through this County in 1805-6, his Mother being a Flat Head Indian woman. From the marked difference in his appearance to that of his swarthy companions, he might well have been taken for a white man. He was totally unable to speak the English language.[75]

Joseph AEneas was with this party, and [he was] said to be by Captain John Mullen, in his Military road Report, an

---

74. For the history of the Jesuits' Salish (Flathead) mission effort, centered at St. Mary's Mission in the Bitterroot valley, see Johnson, *Flathead and Kootenay*, 255-83; Robert Ignatius Burns, S. J., *The Jesuits and the Indian Wars of the Northwest* (New Haven: Yale University Press, 1966), 42-60 et passim; and John Fahey, *The Flathead Indians* (Norman: University of Oklahoma Press, 1974), esp. 64-88.

75. Fahey, *Flathead Indians*, 318, n. 21, mentions the tradition that Clark had a son by a Salish woman. This man his tribe's people later introduced to whites as Peter Clark. For Clark's account of his meeting with the Salish, which makes no mention of any personal relationship, see Moulton, ed., *Journals of the Lewis & Clark Expedition*, 5:187-88.

Iroquois Indian. AEneas, who speaks good English, informed me that he had lived when young in the lower part of Upper California, of which country I understood him to be a native. He was a man of unquestioned bravery, but with the strong Indian instincts of blood thirstiness and cruelty to the Snakes or Shoshones & Bannacks, hereditary enemies of the Flat Head Nation. It was said, and by himself claimed, that he had killed and scalped some twenty Snake Indians during his life time, and he boasted that that tribe could never get even with him by taking his scalp, he was so far ahead of them. His last exploit had occurred only a season or two previous in the upper part of Deer Lodge Valley.

Learning that a Snake Lodge had been struck there, AEneas started off alone to capture its occupant and add another to the scalp trophies that adorned his belt. Mounted on horseback, he commenced circling round the Lodge, drawing nearer and nearer until his war victim, who was intently watching the movements, frought to him with the issues of life or death, stepped from the lodge, leveled his rifle, and sent a bullet whizzing through the air at his approaching enemy, but which failed its intended mark. AEneas jumped from his horse, the Indians rapidly approached each other, and although the Snake Indian was terribly wounded by the Colt's Revolving Pistol AEneas carried, they met in a terrific hand to hand encounter, in which finally the Snake succumbed and yielded up his life. "He was the hardest one to kill I ever met," says AEneas to Jack Mendenhall in telling him the story, and to whom I am indebted for it, "and he shouldn't [have] gone so far from home alone, if [he] had wanted to keep his scalp."

The Snake Indian had two women with him, whom AEneas captured, one of whom he sold to Jim Stuart, and was kept by him as wife or mistress. The other was disposed of in like manner to some of the Mountain Men in the vicinity.[76]

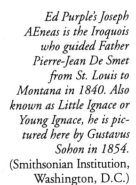

Ed Purple's Joseph AEneas is the Iroquois who guided Father Pierre-Jean De Smet from St. Louis to Montana in 1840. Also known as Little Ignace or Young Ignace, he is pictured here by Gustavus Sohon in 1854. (Smithsonian Institution, Washington, D.C.)

I remained at my cabin till late in November alone with my dog Caesar, a large half breed Mastiff & Hound, which Jack had left with me, and a kitten I obtained from Bob Dempsey, then living below on the Hell Gate.[77] As yet we had little snow, the middle of the day was warm and pleasant, and I employed myself in hunting and fishing, which latter was about as good Sport as earlier in the season. For bait I used Muskrat livers, numbers of which I killed and hung up at the end of the Cabin, to freeze and keep for that purpose.

Tisdale had taken with him to Bannack a wagon load of provisions when he and Sam went over, and to see what I should do with the balance we had, I started over to Bannack

---

76. In describing the same events, James Stuart states that AEneas and his companion, Narcisses, had taken only one of the Shoshone's three wives, the other two making their escape. According to Stuart, the woman was "fair with red cheeks and brown hair and eyes and is evidently half white." For a summary of the story, ending with a rather offhand mention of his de facto marriage, see Stuart, *Forty Years*, 1:197-98.

77. Granville Stuart recorded one incident that interrupted Purple's solitary tranquility in early November, involving the pursuit and capture of an Indian who had taken an ox from one of his neighbors. Stuart, *Forty Years*, 1:228.

about the 20th Nov., taking my Dog along and leaving Puss with Fred Burr. Arriving there, I made arrangements to have Sam and Tisdale build a house, while I returned with the wagon they brought over to Gold Creek for all our things, bag and baggage, with the intention of removing to Bannack. To assist me I hired Jim Conlan and John Sayes, the latter of whom I thought had been born tired, he was so exceedingly lazy.

CHAPTER FIVE

❧❦❧

Gold Creek to Bannack City
December 16 to December 31, 1862

It was not until the 16th of Dec. 1862 that I was ready to start with my loaded wagons, Dog and Cat from Gold Creek for Bannack. A web foot, a name by which Oregonians were known, was at Gold Creek, and urged me to allow him to travel with the waggons to Bannack, and as he pretended to be a good Bull whacker (a term used for ox driver) I consented. What other name beside "Red" he possessed I never knew. There was some snow on the ground when we left Gold Creek, and the ice in the River prevented our taking the Valley road.

We went up the Mullen Road west of the creek sticking into the Deer Lodge Valley 5 or 6 miles below Johnny Grant's. The second day out "Red" overturned one of the wagons, burst open the sacks of flour, and strewed the plain with a "wreck of matter," nearly killing Puss, who was riding on the load, and for whom I claim the honor of being the first of her species in Bannack City.

The third day (19th Dec.) we had a fall of snow a foot deep, which increased the difficulty of travelling. Our progress was so slow that on Christmas eve we had only reached the First Crossing of Deer Lodge [Creek] at the head of the valley.

Ed Purple's brother, Dr. Samuel Smith Purple, a respected New York physician, was pictured (p. 4) in the book about his brother, In Memorium: Edwin R. Purple, *that he published in 1897.*

During that day I met Tom Lavate on his way back from Salt Lake City, with several letters for me, among which were two from my mother and Brother Samuel, the first news I had from home since my arrival in the Country. Lavate had left Deer Lodge late in November to escort a party to Salt Lake, who feared that the deep snows then expected to fall would cause them to lose their way, to prevent which he was taken as a guide.

On Christmas day Conlon broke down his wagon, to repair which we were delayed an entire day. On reaching Big Hole the crossing was very bad, owing to the thick heavy ice which had formed on each side of the river, the current being open, rapid, and at least two and a half feet deep in the center of the stream. In getting over, the mess box containing our Camping utensils fell from "Red's" wagon, and spoons, Knifes & Forks, Tin plates and Coffee mill & Pots strewed the bottom of the River. We could not get along without these necessary articles, and building a rousing fire near the water's edge, "Red" unstripped and waded in to fish them out. It was bitter cold and he worked like a Beaver, securing most of the things—the most valuable service he performed on the trip. His teeth chattered like a red squirrel's in harvest time, and so benumbed was he, I think he would have frozen had we not assisted in dressing him.

This journey, owing to the weather and bad condition of the roads, was prolonged and harassing, and when on the evening of 31st of Dec. we reached Bannack City, I was heartily glad that it was finished.

Bannack in the 1860s (Montana Historical Society Photograph Archives)

CHAPTER SIX

⚜

Bannack City
December 31, 1862, to February 1863

Tisdale and Sam had completed the house, and we unloaded our things that night under its roof. Tom Pitt,
Morrell, and Jack Mendenhall had some time before returned
from Salt Lake with Flour & other provisions for the Miners,
and I found myself surrounded by numbers of old acquaintances from Limhi, Deer Lodge & Gold Creek.

With a retrospective view of the history of Bannack and
vicinity on Willard's Creek, from the time John White's party
first Commenced Mining after our Meeting on the Divide
Creek Aug. 3d, I shall allude to such events as came directly
under my notice from this time, during my stay in this
region. As this record is of a private character, I shall mention
names, circumstances & characters with a freedom, notwithstanding their truthfulness, which would hardly warrant their
being made [known] in a public manner.

Gold was first discovered on Willard's Creek by John
White & company on the 4th day of August, 1862, at a point
some four miles below the present site of Bannack city, in
what is now (1863) known as White's District. White's party
numbered seven men, among whom was one Raymond who had
been living some three years in the vicinity of Deer Lodge. Vague
stories were told among the Indians of a shining metal which was
to be found on this creek, and as the explorations of Gold Tom,

and the mines of Gold Creek then being worked, had proven the existence of the precious metal in this country, White accompanied with Raymond and party determined to prospect the Sources of the Beaver Head River.

Immediately after the discovery, a party from Fort Limhi bound for Deer Lodge followed down the creek, and prospecting the bluff opposite the lower end of the present townsite of Bannack, on the 10th of August struck gold in large paying quantities. Of this party was Harry Hensted, principal Butcher & Baker in Bannack in the winter of 1862–3. The news of these discoveries rapidly spread over the country, and the few people in it hastened to the new diggins. Nothing had yet been found that would compare with them in richness and extent.

The Gold Creek, North Boulder, Prickly Pear & Big Hole mines were soon deserted, and the far off Oregon diggins lost their power of attraction.

The site selected for the town was on the north side of the Creek, where it enters the hills on a bench of ground which rises about 15 feet from the Creek bottom and varies from 1 to 200 yards in width being some 600 yards in length.

(The hills at the edge of which Bannack is situated extend in broken and detached peaks from Horse Prairie's Creek in the south to the Bald Mountain Range in the north, a distance of 40 to 50 miles. They are irregular, and disconnected from the Ranges of Mountains that divide the waters of the Columbia & Missouri. Their greatest width is through where Willard's Creek finds its passage, from which point they narrow up as they extend northward towards the source of the Rattlesnake. Most of these hills are bare, some rocky, but a few are well timbered with mountain pine.)

(They render the town difficult of approach from its eastern side with loaded wagons. The Road leading from Salt Lake City by the way of Horse Prairie is comparatively good.)

The cabins were at first built in two rows small in size, low and close together, with a wide space in front, the intention being to form a Block House or hollow square for defence against the Indians. The number of people who flocked to the place led them to abandon this design, which gave to the town the increased width its only street possesses at the lower end.

The ground was covered with a heavy growth of large brush. Clouds of Grasshoppers filled the air and swarmed on the hills and Creek valley, which circumstance gave to the stream the name of Grasshopper Creek. Few at that time were aware that Lewis & Clark, the first white men who saw it, had given it the name of Willard Creek, after one of the members of their expedition.[78]

This creek is fed by snow and springs from the east side of the range of Mountains in which it rises, and whose western side sheds water which flows into the Columbia River. Its sources are less than a mile from those of the latter stream.

From the base of the Mountain it flows some thirty miles through a grassy plain destitute of trees, but through which its meanderings can be traced at a distance by the willow and alder bushes that line its banks, until it reached Bannack City. Here it enters the hills, which joining more closely together a mile and a half below, from a narrow and rocky Canyon, through which it passes for two or three miles, then breaks from its confinement as the hills slightly widen, forming successive small flats or bars through which it flows, till it empties itself 20 miles below Bannack into the Beaver Head or Jefferson Fork of the Missouri River.

The town was called Bannack after the Indian tribe in whose country it was located. The town proper is built on a bench of ground which rises some fifteen feet from the low

---

78. For this circumstance, see Moulton, ed., *Journals of the Lewis & Clark Expedition*, 5:98.

Robert Halliday drew this picture of Bannack in 1862 when he visited the Montana goldfields with the federally sponsored Fisk expedition. (Montana Historical Society Museum)

flat through which the creek runs. This high level [bench] varies from one to two hundred yards in width, and is some 600 yards in length. A number of small hills rises north of the town, to which the plateau extends, and which forms a narrow valley forty or fifty yards in width, and extending back from the centre of the town a half mile or more. A dry rocky bed runs through this little valley, only filled with water by melting snows in the spring. A gallows erected here in 1863 gives this little valley the name of Hangman's Gulch.

Across the creek and southwest from Bannack is a flat, somewhat higher than the one opposite the town, which was taken possession of principally by Minnesota Men and their families, upon which they erected their Log cabins, and was immediately dubbed Yankee Flat, though really a part of Bannack.[79]

By a singular coincidence, at the same time Gold was found here, it was also discovered in the Boise Country, 500 miles away to the west, and the settlement named Bannack City, it being also in the Bannack Indian Territory. In March following both towns were embraced in the boundaries of Idaho Territory, then just organized, and much Confusion was thereby occasioned in the transmission of letters, to avoid which our town was called East Bannack. Upon the organization of Montana in 1864 the original name of Bannack City was resumed.

The Boise Mines were discovered in August 1862 by a party of prospectors from Oregon. Their number being small, they were attacked by Indians. One of their number named Grimes was killed, and the party compelled to abandon their mining operations and flee for their lives out of the country. In the meantime the immigration that had gone down Snake River from Meek's and Richard's Ferry reached the new diggins. These were followed closely by others who went back from Fort Limhi, upon finding there was no gold at that place. Grimes' party communicated their discovery on their return to the Oregon settlement, and a large number of persons accompanied them back to the Boise Basin, so that by the 1st of Dec. 1862, there was a larger population by far in these diggins than at East Bannack. These are also known as the Owyhee Mines, from the name of the Stream that flows into the Snake River near Fort Boise.[80]

Other discoveries [on Willard's Creek] rapidly proved as rich or richer than the first, and cabins soon dotted the banks of the stream below at Marysville, foot of Nugget hill,

79. The "Yankee" Minnesotans were mainly emigrants who had come west with the Fisk expedition the previous summer.

80. For a more detailed account of the Boise Basin discoveries, see Bancroft, *Washington, Idaho, and Montana*, 259-61; and Greever, *Bonanza West*, 262-71.

Jimmy's Bar, mouth of the Canyon, and extending down to White's District.

Probably not half of the inhabitants on the Creek had sufficient provisions to last them through the winter. Many from their long summer's travel were nearly destitute, but a few week's work amply furnished them the means of obtaining more in the upper Mormon Settlements. To these [settlements] twenty or thirty teams were dispatched in October, and before deep snow fell, returned with sufficient Flour & other necessaries to supply the wants of all.

This was the first permanent settlement in the region now known as Montana Territory. Three-fifths of the people who wintered here in 1862–3 were from Colorado, one fifth were Minnesotians, and the balance, making in all 5 to 600 persons, was made up of Oregonians and former Gentile residents of Salt Lake City.[81] (All persons not members of the Mormon church are called Gentiles, at Salt Lake City.)

In the absence of such restraints as the law imposes upon the old communities, it might be supposed that among the people of this new settlement would be found those ready and willing to infringe upon and violate the rights of their fellows, whenever self interest or wicked desire should prompt them to such action. But the history of Bannack City for the first four months of its existence was one of order, peace, and good fellowship among her citizens. The turbulent spirits that had come with the first wave of emigration were too few in

---

81. A comparable estimate appears in James Fergus, "A Leaf from the Diary of James Fergus, Relative to the Fisk Emigration Party of 1862, and Early Mining Life at Bannack, 1863," *Contributions*, 2:152-54. Fergus makes no mention of emigrants from Utah, but points out the presence of "prospectors and roughs from the west side"—including men like Plummer and those of his associates who had lived on both sides of the law in Idaho and California. Consequently, Fergus adds, "there was little harmony, and good men from those three parties took longer to find each other out, to know who were roughs and who were their friends, than if they had all been from one place or longer acquainted."

numbers to mar the general harmony. The work in inaugurating that carnival of crime, highway robbery, assassination, and murder which is a part of the early history of Montana had not yet commenced. A few of those who rose to their bad eminence of infamy afterwards were here, but quiet and reserved, attending to their labour and mingling (with as yet only half formed notions of evil) with the better classes.

As soon as Gold was discovered by White & party they determined what amount of ground should be held by each person as a mining claim, and each of the Mining Districts, as fast as Gold was found along the creek, pursued the same course. The rule was to give to a person making a new discovery two claims. If two or more made the discovery, only one extra claim was allowed to be divided among all of them. Creek claims were Each 50 ft. front on the creek, running back to the base of the hill. Hill claims were 50 feet wide, 100 in length.[82]

In November a general meeting of all the Miners of the Gulch was held at Bannack, for the purpose of electing a President. This person was clothed with the powers of Judge, with or without a Jury as the parties might elect, to decide in all cases where the right to Mining property was involved. His judgment was final, and he could if necessary invoke the power of the whole Gulch to enforce his decrees. E. P. Lewis of Colorado was elected the First President, and held his office till the Mines of Alder Gulch were discovered, when he resigned. W. Emerick was then elected in his place, soon after which the establishment of regular organized Courts of Law at Bannack done away [with] the necessity of this office entirely.

---

82. The original record of placer discovery claims on Grasshopper Creek, dated August 30, 1862, can be found in the Bannack Mining District Records, SC 238, MHS Archives, Helena. Andrew Murray entered the first lode claim on November 12, 1862.

In the month of October 1862, Charles Benson, H. Porter, E. Porter, and C. W. Place discovered a ledge of Gold bearing Quartz rock a mile below Bannack, and some 500 yards north of the creek. They found large deposits of Gold in a free state, which had crumbled from the rock, and mingled with the earth around it. This was the first Gold Quartz Ledge that had been found in the country, and was named the Dakota, after the Territory within whose boundaries it was then located. Following this and immediately after, Wilson Wadams discovered another Ledge, named after himself the Wadams, situated south of [the] Dakota on the high hill across the creek.

Then commenced the furor for hunting Quartz. The finding of the Yorkshire and many others rapidly followed. Andrew Murray at a point six hundred feet west from the Dakota Discovery struck a vein of Gold Quartz which exceeded in richness anything the oldest miners in the Country had ever seen. Murray would frequently obtain $75 to $100 of Gold from a small sack of the earth and rock, which at night he would carry from his claim to his cabin, pound up in a wooden mortar, and wash out. Whenever a new discovery of Quartz was made, the most intense excitement was manifested by everybody. Men might be seen running as though their lives depended upon reaching the point, across the creek, up the steep hills, panting for breath, never stopping an instant, on they go, with two wooden stakes ready to stick in the ground to mark the bounds of a Quartz claim they believed was to give them possession of millions. Of this they never tired; and every day in the week if occasion presented, they done the same thing over.

From none of these was any considerable amount of Gold obtained except the Dakota. The yield from the Dakota Discovery from March 8th to Oct. 3d, 1863, was $7,054.36, over and above all the expense of labour, tools &c. in working

Wilson Wadams, seen here much later in life, arrived in Bannack on September 18, 1862, with his wife Mary Jane. (Beaverhead County Museum Association, Dillon)

it. The work done during this time was not constant, and in April, May and June there was nothing whatever done upon the mine. It was generally believed that most of these Ledges would pay well, when suitable machinery could be obtained to crush the rock & Extract the Gold.

In the last week of October, 1862, died Wm. H. Bell, a native of Ohio, of Mountain fever, a disease he contracted after his arrival here. He was apparently a stout, rugged man, and was sick only a week. He was the first man that died at Bannack.[83] In November Geo. Ives and Geo. Carhart, who were living at Gold Creek, came to Bannack on a gambling tour, and a day or two after their arrival got into a wordy dispute near the Bakery, which ended in the exchange of several shots, one of which from Carhart's pistol took effect in Ives' body. From this wound Ives soon recovered, the difficulty was adjusted, and the two men became fast friends. This was the first shooting scrape in the town.[84]

On November one Charles Guy left Bannack with his ox team to procure a load of flour at Box Elder, the nearest Mormon settlement. A few days after, his body was found horribly mutilated, on Red Rock Creek, and it was generally supposed he was killed by the Indians.[85]

---

83. The actual date of William Bell's death, described by Nathaniel Langford as the first death at Bannack City by natural causes, was November 2, 1862. Entry of November 3, 1862, James Henry Morley, "Diary of James Henry Morley in Montana, 1862–1865," typed manuscript, MHS Library, Helena. Since Bell had made a death-bed request to be buried with Masonic ceremonies, as Langford recounts in detail, his funeral services the following day brought together the many members of the Masonic order who were present at Bannack that winter. Langford conducted the funeral ceremonies and then took a leading role in organizing a Masonic lodge at Bannack. Later he claimed that the Montana vigilante movement, which he typified as a movement of retribution against a band of desperadoes, "had its origin in principles traceable to the stalwart morality which is ever the offspring of Masonic and religious institutions": Nathaniel Pitt Langford, *Vigilante Days and Ways: The Pioneers of the Rockies* (New York: A. C. McClurg & Co., 1912), 181-85. The claim that the Montana vigilante organization was mainly a conspiracy of Masons is examined in detail and refuted in Merrill G. Burlingame, "Montana's Righteous Hangmen: A Reconsideration," *Montana The Magazine of Western History*, 28 (October 1978), 36-49; but see also J. W. Smurr, "Afterthoughts on the Vigilantes," *Montana The Magazine of Western History*, 8 (April 1958), 8-20.

84. With some variation in detail, Thomas J. Dimsdale, *The Vigilantes of Montana; or Popular Justice in the Rocky Mountains* (1882; reprint, Norman: University of Oklahoma Press, 1953), 30-31, and Langford, *Vigilante Days and Ways*, 133-34, each provide accounts of this fight between George Ives and George Carhart. Only Purple dates the episode (however questionable his dating may be) and identifies it as the first "shooting scrape" in Bannack's history. Carhart's name is variously spelled. Dimsdale puts it down as "Carrhart" and Langford follows that precedent. Both in the 1862–1863 Bannack City census listing and in his manuscript Purple adopted the "Carhart" spelling. George's brother John, who arrived in Bannack from California after George's death and subsequently settled at Dillon, spelled the family name as "Carhartt," according to an obituary notice that appeared in the *Dillon Examiner*, December 25, 1907. George Carhart, identified as a member of the Whig Party, had represented Colusa County in the California state legislature's 1853 fourth session. James D. Driscoll and Parry R. White, comp., *List of Constitutional Officers, Congressional Representatives, Members of the California State Legislature and Members of the Supreme Court, 1849–1985* (Sacramento: State of California, undated), 77.

85. Neither Dimsdale nor Langford record the death of Charles Guy, but it is noted in Stuart, *Forty Years*, 1:234 and n. 78.

The weather during the fore part of the winter was clear and cold. The labour of mining was carried on with difficulty. The pay dirt from the claims was heaped up in piles, upon which fires were built, and the water used in washing the earth was heated for that purpose. Upon to the 1st of Jany. 1863 little snow had fallen, but on that day a snowstorm which commenced during the night covered the earth with thirty inches of snow. This suspended all mining operation except underground work for two or three weeks.

At this time the public places of resort were the saloons of Bill Goodrich, Durant & Perie, and Jack Gunn's, where lighting rod whiskey was sold at 25 cents per glass in gold dust. The demand for this article being great, the supply limited, the business soon played out, there being no way of replenishing the stock in trade. I purchased the remnant of Merchandise Jack Mendenhall [had] brought from Salt Lake, which reduced the stores in town to two, Wm. Butts' and my own. A meat market and Baker was established by Harry Hensted and Andrew Luje. Henry Crawford also engaged in Butchering cattle about the same time.[86]

The Bake shop was a great resort for Pie Eaters, where they indulged their taste alike for Seven up, Euchre and dried Apple pies. Here could be seen from early morn till midnight George Hilderman, the Great American Pie Biter, a man not unknown to fame afterwards in Montana. On a wager he agreed to bite through the thickness of a half dozen pies, and his capacious mouth would have permitted the performance of the feat and his winning the bet, had not his opponent wrung in a tin plate between the stacked up pile, against which he essayed his teeth in vain. He lost his money by a

---

86. For a general account of Bannack City's early business and social life, see F. Lee Graves, *Bannack: Cradle of Montana* (Helena: American & World Geographic Publishing, 1991).

foul, but retained his reputation as the Champion Pie Eater of Bannack. His love of Pies was wonderful.[87]

Among the trades represented were shoemakers, Tailors, Blacksmiths & wheelwrights, and one Barber who found employment principally in the Hair Cutting department of his profession in Bill Goodrich's saloon.

On the 2d day of Jany. [1863], George Edwards, a teamster employed by Wm. Butts to bring the stock of merchandize purchased by him of Capt. Wall at [Little] Black Foot to Bannack, left the town for the purpose of hunting up and looking after his Cattle. He started up the creek through the deep snow, and was never seen afterwards alive. After two days had elapsed search was made by Budd McAdow and others of his friends, who feared he had lost his way and frozen to death in the snow. A week after a man by the name of Duke, a partner of Jimmy Spence, who was herding Cattle up the creek, found Edwards' clothes done up in a bundle, perforated with bullets, frozen and bloody, and thrust into a Badger's hole. This made it apparent he had been murdered by white men, as had Indians done the work they would have taken the dead man's clothing. It was suspected by some that a man by the name of Cunningham [Cleveland] had committed the deed, for the purpose of robbery, as Edwards had $130 on his person New Year's day, while Cunningham, who was dead broke before the murder, was flush immediately after. Whatever became of the Horse and saddle Edwards was on the day he left town always remained a mystery.[88]

---

87. The published version of Purple and Sander's 1862–1863 Montana census listing shows this name as George Hillerman, but both Dimsdale and Langford also call him Hilderman. In nearly identical terms, Dimsdale and Langford each describe him as a rather weak, "somewhat imbecile" old man. Langford also relates the pie-biting story, which may have been introduced in Hillerman/Hilderman's defense when he was tried by the vigilantes as a member of Plummer's gang. He was found guilty, but the court exercised leniency in passing a sentence of banishment. Dimsdale, *Vigilantes of Montana*, 116–18; Langford, *Vigilante Days and Ways*, 307-8.

*Perry W. "Bud" McAdow,
pictured here when he served as a
delegate to the Montana
Constitutional Convention in
July 1889, mined with the Stuart
brothers on Pioneer Creek, a
tributary to Gold Creek.*
(Montana Historical Society
Photograph Archives)

The death of Edwards was the first of a long series of
cruel, cold blooded, fiendish murders, coupled generally with
highway robbery, that shed a gloom and terror over the
country for more than a year, and carried sorrow and mourning to many a distant fireside. It required something more
than the murder of one or two individuals to awaken this
young Community to a realizing sense of the insecurity of life
and property, and the desperate character of the few ruffians,
robbers & murderers who had taken up their residence
among them. These were mostly from the "other side of the
Mountains," men whose crimes had driven them, unwhipt of

---

88. At this point memory apparently tricked Purple; other authorities give the
name of this man, the suspected killer who soon after became one of Henry
Plummer's victims, as Jack Cleveland. Dimsdale, *Vigilantes of Montana*, 26-29;
Langford, *Vigilante Days and Ways*, 24, 66-68, 131-33. Moreover, the Purple and
Sanders 1862–1863 Montana census listing includes the name Jack Cleveland,
with a notation that he was "Killed by road agent Plummer, December 1862."
Contributions, 1:336. Because Purple writes that Edwards left Bannack City in
early January (while Dimsdale claims the time was February 1863), and because
Plummer killed Cleveland some time later (Langford states "a few days after the
disappearance of Edwards"), the discrepancy in dates cannot be reconciled; yet
there is general agreement among all sources regarding the sequence of events.
See also Stuart, *Forty Years*, 1:234-35 and n. 78.

justice, from California and Oregon, whose hands dripped with human blood, and on whose souls rested the awful curse and sin of Murder. It was not until Jany. 1863 these men ceased their wanderings to and fro through the Mountains, and settled down to live in Bannack. When they first reached here, their ostensible business was the same as all newcomers, that of Gold digging. They added to this occupation, to which they really paid little attention, that of sporting men or Gambling, a habit almost universally prevalent among western men, and more particularly so in new mining Communities.

Of the most noted of this class, and who finally suffered death for the crimes they committed, were Henry Plummer, Geo. Ives, Hays Lyon, Wm. Graves alias Whiskey Bill, Wm. Hunter, Cyrus Skinner, Buck Stinson, Ed [Ned] Ray who came in the spring, Wm. Moore & Charles Reeves. The last two escaped from the country, and were heard of last in Mexico. Jack Cunningham [Cleveland], Charles Forbes, alias Ed Richardson, and George Carhart met violent deaths at the hands of their Companions in deeds of bloodshed and Murder.[89]

A month after the death of Edwards, Henry Plummer killed Jack Cunningham [Cleveland] under the following Circumstances. They were companions and came from Lewiston, or the Mines at Oro Fina (Spanish words meaning Fine Gold) in Oregon together, to Gold Creek in the month of October. They had been travelling about the country

89. This listing is far from a complete roster of the men killed or banished by the Montana vigilantes for their crimes in the winter of 1863–1864. For other listings see Sanders, ed., *X. Biedler: Vigilante*, 85; Dimsdale, *Vigilantes of Montana*, 23-24; Langford, *Vigilante Days and Ways*, 315; and Graves, *Bannack*, 41. A comparison of the various older lists, with critical comment on the discrepancies, can be found in Mather and Boswell, *Vigilante Victims*, 165-67 et passim.

ostensibly hunting for Gold Diggins. Plummer and Cunningham were both at Bannack in November, but remained only a short time, leaving for Sun River, where a Government [Indian] farm was located and then in charge of a Mr. [James] Vail, who with his family, including his wife's sister, a Miss Electa Bryan, resided. Plummer remained there or in the vicinity, while Cunningham went back to Bannack, was there two or three days after Edwards was murdered, and then left without exciting more than a mere suspicion that he was the Murderer. He then returned with Plummer, reaching Bannack during the last days of Jany.

A day or two after their arrival, early in the afternoon, Jack entered Bill Goodrich's saloon. At the time Plummer, Bill Moore, Jeff Purkins, Goodrich, and one or two more were in the saloon, some of whom were engaged in a game of cards. Cunningham, who was full of whiskey, was irritable and disposed to pick a quarrel with Purkins, who acted sometimes as Bar Keeper for Goodrich. He alledged that Purkins had cheated him, or owed him money which he demanded in the most insolent and insulting language. Purkins, thoroughly intimidated, as well he might be, got out of the place and went off to arm himself.

In the meantime Cunningham continued his boisterous conduct, exclaiming in a loud tone that he was the "Big Bamboo chief" and "Knew every cutthroat son of a bitch in the Rocky Mountains." The attempts made to quiet him were unavailing, until Plummer quietly rose from his seat, and seizing a pistol which Moore handed to him behind his back, approached Cunningham and said "I've got tired of this dam nonsense," pointed his pistol, and discharged two shots into his body. Cunningham was too drunk to defend himself, and in the awkward attempt he made to draw his pistol, fell onto his knees. In this position , half reclining, he says, "you won't shoot a man when he is down?" "No" says Plummer, "get up you son of a b—h," and holding the muzzle of his pistol to his face so near [that] the powder burned it, he fired a third shot, followed immediately by a fourth through the head, which, singular to tell so tenacious of life was Cunningham, it did not prostrate him, but he still retained his reclining position.

Plummer then walked to the door of the saloon, and as he stepped out waved his Pistol round his head and exclaimed, "My name is Henry Plummer." It was not a minute from the time Plummer took the pistol from Moore, till he was out-doors walking down the street.

This Murder occurred nearly opposite my store, across the street, and after the firing had ceased I stepped over to see what was the matter. As I left the store Plummer stepped from the door and walked away. I found Cunningham resting on his elbow, and cursing most piteously for "some son of a b—h" to kill him, and heaping upon himself the vilest epi-thets that I ever heard fall from a man's lips. Henry Crawford, coming in a short time after, with the assistance of some of the bystanders conveyed him to his Butcher shop, where in about four hours he died.

While lying at Crawford's shop he was visited by George Ives, who he requested to take his six shooter and blow out his brains, to relieve him from his suffering. Ives' reply exhibits how abandoned, how lost to every kind emotion his heart was. He says, "Jack you'll be in hell soon enough; die like a man, and don't make such a d—m fuss about it."

Before Cunningham was dead, Plummer, Reeves, Moore and Ives came into my store, apparently as unconcerned as though nothing unusual had occurred, and the only allusion that was made to the circumstance, was that when one of them casually asked if Cunningham was yet alive, Plummer remarked that if he thought so, "he would go and finish the Son of a b—h." They then stood up by a Post that supported the roof of the store and measured their heights, the marks of which could be seen a long time after.[90]

Plummer had little of the bragadocio, or slashing style of manner that characterized the behaviour of his comrades. He was a man about 28 years of age, 5 feet nine inches in height, and would weigh about 135 lbs., brown hair, light eyes, except when angry—when they grew black and glistened like a rattlesnake's. He spoke in a low, quiet tone of voice, a habit which never deserted him, even when labouring under such intense excitement as the Murdering of a human being must have produced. When drunk he had a practice of grinding his teeth, which those well acquainted with him understood to be a warning to those with whom he was not on friendly terms. He was scrupulously neat in his person, generally wore an elegant fitting frock coat of black broadcloth, and was the best dressed man in town. He had the reputation of being the quickest, in drawing and discharging his Six shooter, of any man in the Mountains except Charley Forbes, who in the rapidity and accuracy with which he fired a pistol was not excelled by any one then in the Country.[91]

Plummer was originally from Lowell, Massachusetts, where before going to California he supported a widowed

---

90. Despite the name confusion, Purple's narrative provides fuller, more specific details of this affair and especially the circumstances of Cunningham/ Cleveland's death than other sources. See Dimsdale, *Vigilantes of Montana*, 27-29; and Langford, *Vigilante Days and Ways*, 131-33.

91. Frank Thompson, who spent considerably more time with Plummer on terms of familiarity than did Purple, in his autobiography describes him as five feet ten inches in height and weighing 150 pounds. Thompson, "Reminiscences," *Massachusetts Magazine*, 6 (October 1913), 161. In their extended critical reexamination of Henry Plummer's career, R. E. Mather and F. E. Boswell add an additional inch to his height, declaring without citing a source that "He looked to be about 5'11", and to weigh around 150 pounds." R. E. Mather and F. E. Boswell, *Hanging the Sheriff: A Biography of Henry Plummer* (Salt Lake City: University of Utah Press, 1987), 19-20. Langford's description of Plummer, dwelling on his character rather than his physical appearance, appears in *Vigilante Days and Ways*, 130.

mother, brother and sister by keeping a thread and needle store.[92] I shall have occasion to refer to him again, and for the present will not go further into the history of his previous life.

The murder of Cunningham attracted but little attention, and in a few days was comparatively forgotten. The general feeling seemed to be, without much inquiry into the causes that led to it, that a bad man, perhaps the worst in the Community, had been got rid of, and if others could be disposed [of] in the same way, the town would be the gainer thereby.

If there had existed any desire to punish Plummer, no single individual would have been willing to have put himself forward in such a movement, and invoked upon his head the vengeance of Plummer's particular friends, the number of which, or who they all were, at this time was simply a matter of conjecture. As yet the people of the town were in a great measure unaquainted with each other. No one man could claim a half dozen reliable friends, upon whom he could depend in the event of his taking steps for the punishment of crime. The feeling was a deepseated one among the better class that largely predominated, that anything done in that direction, no matter by whom, was frought with danger, and at the peril of one's life.

In short time the current of events resumed their quiet and peaceful flow. Balls and dancing parties were held two and three times a week, while a Dancing school was open every night except Sunday, under the supervision of Buzz Cavan and his partner, S. P. [Lou] Smith.[93] These parties were pleasant and agreeable, and patronised as they were by all the Ladies, married and single, done more to cultivate a spirit of

---

92. In fact, as Mather and Boswell have established through an examination of court records for Nevada County, California, Plummer's birthplace was Addison, Maine, where his family had been established for three generations and where he lived until leaving for California in 1852 at age 19. Mather and Boswell, *Hanging the Sheriff*, 195-99.

good will and friendly feeling among the citizens than all other means combined. They were held in a large hall built by J. M. Castner, of split logs, with a smooth floor adapted for these purposes.[94] Order and decorum were strictly observed by all, and it never happened any time that these social gatherings took place, during the entire winter, that a rude or improper word was spoken, or any breach of the ordinary rules of civility occurred. Cavin & Smith were capital violinists, and there was no grumbling at the Music. The suppers were generally gotten up under the supervision of Mr. & Mrs. Jas. Harby, Mr. & Mrs. Gotfried [Godfrey], and H. F. Morrill [Morrell], who was famous for making good coffee.

The Belles of the town were Miss Mary Donnelly, whose style of beauty was of the blonde order, and Miss Matilda Dalton, a charming brunette. Miss Dalton was nicknamed Desdemona, and a select few whom she permitted to worship at the shrine of her beauty, familiarly called her Dez. She was a tall, magnificently formed woman, about 22 years of age, dark hair and eyes, regular features, and would turn the scales at 155 pounds. One of the boys said he thought all she eat went down into her feet, they were so large; but he was a fellow that had received his walking papers from her, and couldn't call her Dez any more. It was said that poor P. C. Woods blowed out his brains because he could not marry her. After the settlement of Virginia City she removed there and

---

93. For the active role of Buzz Cavan and Lou Smith in the social life of Bannack City, see Stuart, *Forty Years*, 1:267-68; and Great Falls *Tribune*, October 2, 1949 (copy in the Bannack City Vertical File, MHS Library). For another account of Bannack City dances, vividly recalled by a woman who in 1863 was attending her first dance at age fourteen, see Sarah Wadams Howard, "Reminiscences of a Pioneer."

94. James M. Castner, born in Maine in 1805, was a pioneer in the hotel business at St. Paul, Minnesota, then at Bannack City, and subsequently at Virginia City, where he was elected mayor. He died in Bozeman in 1876. (Bozeman) *Avant Courier*, April 28, 1876; Helena *Herald*, April 26, 1876.

married a French man by the name of Thibalt, with whom she shortly after left the Territory.[95]

Of Miss Donnelly I shall speak hereafter. On the 22d day of Feb'y. a grand Washington birthday ball was given by Mr. & Mrs. Henry Zoller, at Yankee Flat, at which there was a general turnout of the people of both places. It was a pleasant reunion, and dancing was kept up as usual till broad daylight.[96]

One of the most important events that transpired after the Discovery of Gold on Willard's Creek, and which aided materially the prompt and early settlement of the Country, was the battle of Bear River, fought by the California Volunteers under Col. Edward P. [Patrick E.] Conner against the Snake and Bannack Indians. This battle took place some fifteen or twenty miles above the old California Crossing on Bear River, in the month of Jan'y. 1863. To be attacked in the dead of winter by U.S. soldiers was a plan of waging war to which the Indians were unaccustomed, and which they did not believe was likely to occur.

A band of about Three Hundred of these savages, after their usual summer depredations on the plains, had taken up

95. While no available sources provide biographical information on Mary Donnelly, Matilda Dalton's manuscript autobiography, "History of a Montana Pioneer," in the holdings of the Montana Historical Society, is printed in Mable Ovitt, *Golden Treasure* (Helena: State Publishing Co, 1954), 253-54. An extended obituary appears in the Great Falls *Tribune*, February 7, 1932 (copy in the Thibadeau Vertical File, MHS Library). A native of Maine, born in 1843, Dalton came to Montana with her parents from Wisconsin as part of the 1862 Fisk expedition. The family located in Bannack City in December of 1862, then moved to Virginia City a year later, where Ms. Dalton's parents both "took the fever" and died within two weeks of each other in January 1864. She subsequently married Zebulon B. Thibadeau and the couple relocated to Oshkosh, Wisconsin, only to return west and settle in Wallace, Idaho, in 1874. Granville Stuart, who fancied himself an expert at nicknames, claimed responsibility for fastening the "Desdemona" title on young Ms. Dalton "because she was beautiful and so good." Stuart, *Forty Years*, 1:261.

96. Mr. and Mrs. Henry Zoller and Miss Emma Zoller appear in the 1862–1863 census listing for Bannack City, but otherwise have remained unknown to history. *Contributions*, 1:342-43.

Matilda Dalton was nineteen when she came to Bannack with her parents in December 1862. A year later they moved to Virginia City where this photograph was probably taken in about 1864. (Montana Historical Society Photograph Archives)

their winter quarters in a ravine above the Bear River Canyon, and in such a commanding position that they deemed it impregnable to any assault that could be made against them. So safe did they regard this retreat, they sent word to Col. Conner, then at Camp Douglas near Salt Lake City, where they were and actually invited the contest. He started with three or four Companies, principally of Infantry, and reached the encampment on a bitter cold day, the 29th of Jan'y., 1863, previous to which the men were compelled to wade the river, after doing which many of them had their hands and feet badly frozen. The Indians had displayed considerable generalship in digging rifle pits and breastworks, and adding to the natural strength of their position in every way their Savage military skill suggested.

Conner in reconnoitering their stronghold found that his men would be exposed to a raking fire from the Indians, from every quarter in which he attempted to approach, and he therefore determined to receive their fire and charge down upon them with his infantry, reserving his mounted men to cut off their flight when routed. When the Bugler sounded the order to charge from the top of the hill, the Indians pierced him dead with their bullets. The Indians kept up a terrible fire as the men came charging down upon them, and held their ground firm and undaunted for a few minutes; but the work soon became too hot and close for them, they broke and fled like deer from the place. Some attempted to cross the river, but were pursued and killed by the soldiers, who used in these close quarters their six shooters and knives, so that out of three hundred warriors it was said that only twelve escaped from the bloody field; and [Connor's command] nearly exterminated the whole tribe.

The loss of the whites was not as heavy as would have been supposed from the circumstances, amounting to less than twenty on the field; but a number died afterwards from their exposure, and several had their limbs amputated, rendered necessary from their frostbitten condition. In the Indian lodges were found numerous articles of women and children's clothing, jewelry, meershaum pipes and other things, evidently the spoils of murdered emigrants' trains, which had fallen into the hands of the savages. A few men not connected with the Army were present and assisted the soldiers in the fight, from one of whom, Mr. Clancy, a reliable man, I had this narration of the battle on the spring following. Col. Conner was promoted to the rank of Brigadier General for gallantry in this fight.[97]

In the latter part of the month of Feb'y., 1863, Charles Reeves and Wm. Moore about Eight o'clock in the evening, after having thoroughly soaked themselves with whiskey for

---

97. For a far less admiring account of the conduct of Colonel Connor and his California volunteers, see Madsen, *The Shoshoni Frontier*, 177-200.

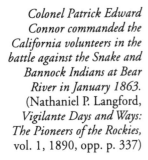

*Colonel Patrick Edward
Connor commanded the
California volunteers in the
battle against the Snake and
Bannock Indians at Bear
River in January 1863.*
(Nathaniel P. Langford,
*Vigilante Days and Ways:
The Pioneers of the Rockies,*
vol. 1, 1890, opp. p. 337)

the hellish and bloody purpose they had in view, crossed the
creek to Yankee Flat, and commenced an indiscriminate
attack upon three or four Lodges of peaceable and friendly
Snake & Bannack Indians. A young Frenchman, Bisette, an
Indian interpreter, standing at the Mouth of one of the
Lodges, and an Indian were killed by the first fire, and a white
man named Woods wounded. They continued firing at
random into the lodges filled with Indian women & children
till the chambers of their six shooters were all discharged,
when they recrossed the creek for the purpose of reloading
their weapons. During this time the Indians were so surprised
at the attack, they Could not believe it was purposely done,
and did not offer any resistance. The Lodges had been pitched
near the foot of the bluff a short distance from the creek, and
during the ten or twelve days they had been there, the Indians
had been civil in their actions, and the whites had treated
them in the most friendly manner. Indeed nothing had
occurred except the disappearance of Edwards the day after
New Year's, to excite any fear or alarm on the part of the
whites towards the Indians.

Bannack miners elected J. F. Hoyt to serve as judge for the miners' court that tried William Moore and Charles Reeves for murder. (Nathaniel P. Langford, *Vigilante Days and Ways: The Pioneers of the Rockies*, vol. 1, 1890, opp. p. 267)

As soon as Reeves & Moore had reloaded their pistols they returned to the cowardly charge and slaughter of the poor defenseless Indians (with a Minnesotian, John [William] Mitchell), and again emptied their six shooters in their midst, killing another male Indian and mortally wounding a squaw and papoose. By this time the rapid firing & the shrieks & the howling of the Indian women mourning for their dead had aroused the people of Yankee Flat & Bannack living near the bluff, who rushed with their arms to the scene of disturbance. To increase the excitement a report was raised that an attack was being made by Indians upon the town. The foremost few on reaching the encampment found the Indians taking down their Lodges and removing up the hill as rapidly as they could. Reeves & Moore were boasting of the revenge they had taken for the murder by the Indians of some of their relations on the plains ten years before; but when some of the citizens denounced their bloody act as cold brutal murder, deserving the hangman's rope, they fled to their cabin in the lower part of town.[98]

---

98. A short account of this incident appears in Dimsdale, *Vigilantes of Montana*, 33-35, while Langford, *Vigilante Days and Ways*, 135-36, provides an even more abbreviated version.

After removing the young Frenchman's body across the creek, the citizens determined to arrest Reeves & Moore & hang them, and had they been found they certainly would have been hung that night. The next morning a posse of 30 or 40 men started in pursuit of the Murderers, who had left during the night in company with Henry Plummer, who had killed Jack Cunningham some time before. They were overtaken in the afternoon of that day on the rattlesnake Creek, 12 miles from Bannack. Reeves & Moore wanted to resist a capture, and would have sold their lives as dearly as possible, had not Plummer advised them to give themselves up. They were brought into town near night and placed under guard, while preparations were made for their trial the next day.

In the morning a large number of the citizens met at Castner's Hall (the largest house in Bannack) and organized a court for their trial. Judge Hoyt, a Minnesotian, was elected Judge, Henry Crawford [elected] sheriff, and George W. Copley [elected] prosecuting attorney.[99] Judge H. P. A. Smith and W. C. Rheem acted as counsel for Reeves, Moore, Mitchell & Plummer. This was the first trial by Jury of a Capital offence in Beaver Head County.

Plummer was tried separately, his offence being in no way connected with the others. It was made to appear by the testimony of the witnesses, introduced in behalf of Plummer, that the killing of Cunningham was in self defence. Great latitude was given in taking the testimony, and Plummer was made a witness in his own case. He claimed to have been a Sheriff in one of the Northern Counties of California, and in

---

99. Of these three elected court officers, biographical information can be found for Judge John F. Hoyt, a member of the Minnesota bar who left St. Paul for the Rocky Mountain gold country in spring 1862. After a year in Bannack City and Virginia City, Hoyt returned to Minnesota; in 1901 he again traveled to Montana to live in retirement with the family of his son, Dr. Mark Hoyt, at Glasgow, where he died in 1905. *Progressive Men*, 1583; Helena *Independent*, October 5, 1905.

discharge of his official duties as such, had with a posse of men attempted to arrest a gang of thieves who had robbed one of Wells Fargo & Co. Coaches of the Treasure Box; that Cunningham under an assumed name was one of the gang. Cunningham had a brother, also one of the gang, who was killed by Plummer in this attempt to arrest the party. Cunningham swore to be revenged and meeting Plummer at Elk City, Idaho, had followed him from place to place until he arrived at Bannack. Here Plummer, fearing he would execute his threat of vengeance, killed him under the circumstances already detailed. The Jury acquitted Plummer, who commenced at once to enlist the Sympathies of the citizens by private appeals on behalf of his comrades.[100]

Mitchell was not regarded as culpable in the murder of the Indians as the other two; and as the trial progressed it became apparent that he would not be convicted, or if so his punishment would be light compared with the others. Judge Smith put forth his best efforts for his clients, and witnesses were produced who swore that Reeves' sister with a party of emigrants bound for California in 1852 were murdered by the Snakes & Bannack Indians on Sublett's Cutoff. This defence, though a meagre one, sufficed to induce the Jury to look with some leniency upon the act, and though they found the prisoners guilty, they adjudged as the punishment that they should leave this part of the Territory, never to return to it again upon pain of death.[101]

100. For Henry Plummer's California career, including his term as the elected Nevada City town marshal, see Mather and Boswell, *Hanging the Sheriff*, 119-72. While serving as marshal, Plummer shot and killed a man under conditions that today would almost certainly be ruled self-defense; but in this case, a jury found him guilty of murder in the second degree. After failing to secure an acquittal on appeal, Plummer spent six months incarcerated at San Quentin state prison before Governor John Weller granted him a pardon. With careful research, Mather and Boswell have constructed the best possible circumstantial case to defend Plummer against the accusations of the Montana vigilantes and their apologists, Thomas Dimsdale and Nathaniel Langford; but none of these authors mentions an earlier clash involving Jack Cleveland and his brother.

The next day after the trial a purse of about one Hundred Dollars was made up among the people of the town for them, and they left Bannack for Deer Lodge valley, in company with John Crow, alias Texas, and a few others of their own ilk and kind.[102] Mitchell remained although sentence of banishment was pronounced against him, and he finally subsided into a quiet and peaceable member of the community.[103]

Charles Reeves, one of the principal actors in this deliberate murder of the Indians, and of the Bisette the Frenchman, was a young fellow of handsome appearance, bright and intelligent, of polished manners and gentle address. He was not over 20 years of age, a native of Pennsylvania, and was far better educated than any of his companions. No one would believe upon meeting him and looking into his frank open countenance that he was the dangerous desperado he was. He had had his man for breakfast before he was 18 years old, had killed one in Oregon before coming to Bannack, and by his cunning and adroitness managed to escape from Montana without punishment. He had the most perfect command of his facial organs, and could assume an innocent expression so natural, joined with a grace of manner and person so Captivating, that he seldom failed to convince a stranger that he was really what he seemed to be—an upright, honest and well disposed person. Among his boon companions, his ribald and

---

101. The most detailed account of this trial and the sentencing of Plummer, Reeves, Moore, and Mitchell comes from the pen of Nathaniel P. Langford, who was prominent in arguing for a jury trial and who served as foreman of the elected jury. Langford, *Vigilante Days and Ways*, 137-47; Dimsdale, *Vigilantes of Montana*, 36-38. See also Mather and Boswell, *Hanging the Sheriff*, 28-29.

102. The (Virginia City) *Montana Post* for November 26, 1864, carried a notice that a man named Crow, alias Preston, had been shot and killed at Pend d'Oreille, after narrowly escaping hanging for stealing horses from the Hudson's Bay Company at Hellgate.

103. In fact, as Purple later relates, Mitchell soon after joined a small party headed for the Salmon River mines, but he and his companions were apparently killed by Indians before reaching their destination.

blasphemous language rivaled theirs, and possessing all the advantages they had in the quickness of sight, and dexterity in the use of the Bowie Knife and Pistol, and the fearlessness with which he engaged in robbery or bloodshed, coupled with his superior education, made him a favorite among them.

Moore was a stout man of considerable muscular power above the average of his fellows, and was more coarse and brutal in his appearance. He and Reeves were fast friends, and worked well together in their schemes of Robbery and Murder.

A few days after the trial, Jack Mendenhall, J. F. Emory, Major Graham and a half dozen others left Bannack intending to go to Fort Benton, and down the river to Omaha and St. Louis with the Mackinaw boats of the American Fur Company. These Boats usually started down the Missouri about the 20th of March, at which time the ice in the River broke up. Mendenhall had some $10 or $12,000 in gold dust with him, the proceeds of his sale of the stock of goods brought from Salt Lake City late in the fall. The other members of the party had an equal or larger amount, which was the first Gold in any quantity that had left the County. Upon their arrival at Deer Lodge they found Moore, Reeves, Crow (or Tex.) and others of the gang, who were somewhat apparently solicitous for the safety of Jack and party, and some of whom proposed to accompany him as a body guard to Fort Benton.

Jack, who always carried a level head, declined their services, thought his party was large enough, all of whom were well armed, and whom he knew to be trusty. He further intimated to Moore, who in company with Hays Lyon and Wm. Hunter had been employed by him as Teamsters in bringing up his goods from Salt Lake—the way they came into the country—that all he had in the world was on the pack mule that carried the party's treasure; that he had deter-

Frank Worden, shown here about 1875, opened a store with partner C. P. Higgins at Hell Gate, four miles east of present-day Missoula, in 1860. When gold was discovered at Gold Creek in 1862, they added a store there, then expanded to Deer Lodge in 1863. (Montana Historical Society Photograph Archives)

mined it should not be taken without he lost his life, whether attacked by robbers or Indians.

The number of the party was increased at Deer Lodge by Frank Worden, merchant at Hell Gate, John Powell, and some others on their annual visit to St. Louis, and arrived safely at Fort Benton, and in April reached Omaha and St. Louis.[104]

---

104. This 1863 trip by Mackinaw boat is not otherwise recorded. Stories of later Mackinaw boat voyages can be found in Langford, *Vigilante Days and Ways*, 492-516; and Mark H. Brown, *The Plainsmen of the Yellowstone: A History of the Yellowstone Basin* (New York: G. P. Putnam's Sons, 1961), 177-82.

Bannack looking north, circa 1875
(Montana Historical Society Photograph Archives)

CHAPTER SEVEN

❧❧❧

Bannack City
March and April, 1863

In March Plummer, who had become greatly exasperated towards Henry Crawford, who had kindly cared for Cunningham in his dying moments, and acted as Sheriff at the trial of Reeves, Moore, Mitchell & Plummer, as already related, made threats that reached Crawford's ears, that he intended to kill him on sight. It was generally believed that Plummer's hatred of Crawford lay in the fact that he thought Cunningham in his dying moments, when alone with Crawford, had divulged some weighty secret of Crime, which he had perpetrated and in which Plummer was also a guilty party. If this was what excited his fears, he could have rested them as unfounded. If there was a secret the dying man had kept it well. The fearful sufferings he endured for four hours, after receiving his death wounds, had extorted no confession of crime, nor even an allusion to his past life. If he had pledged himself to Plummer, not to reveal what lay between them, in the category of their crimes, he redeemed it by the silence of his tongue and lips till touched by death. It was forever safely locked in his breast.

With a heavy conscience of guilt upon his soul and fearing that Crawford had been furnished by the dying man's confession with the clue to some terrible mystery of his life, Plummer resolved to place him in an early and bloody grave.

Whatever his real object was in desiring to kill him, he veiled it completely from public sight. He reported the cause of offence he had against Crawford, to be that he [Crawford] said that he [Plummer] was sleeping with an Indian Squaw called Catherine, the somewhat loose and drunken mistress of one of the old Mountaineers in town.

At this time it was reported that Plummer was engaged to be married, or at least was very sweet on Miss Bryan, the sister-in-law of Mr. Vail at the Sun River [Indian] Farm. This young lady, whom he married the May following and brought to Bannack, belonged to a very religious family, and was herself a girl of the deepest piety. It was said for many months that this was the only family in the country where grace was asked at meals, or the religious custom of family worship observed.

Plummer had first visited the [Indian] Farm during the fall. [He] was a total stranger, but had happened there at a time of day when all of the menfolks were away, the ladies— Mrs. [Martha] Vail and her sister, Miss Bryan—being alone at the house. A short time after his arrival a party of eight or ten Crow Indians made their appearance and demanded food, accompanying the demand by such demonstrations of hostility as frightened the ladies, and alarmed them in the highest degree for their safety. Plummer, observing how greatly their fears were excited, with a boldness of which he was the perfect master stepped up to the principal chief of the Indians and pointing his Pistol at him, ordered him to leave the place immediately. The Indian knew well enough, as he gazed into that Rattlesnake eye, as its flashing brilliancy met his own, that his life would pay the penalty if he disobeyed the low-toned, but no less emphatic order of the young white man; and calling his men hastily left the Farm.

From this circumstance [Miss Bryan] became enamoured of the handsome and courageous stranger, and despite the entreaties and persuasions of her friends, taking counsel alone of her woman's heart, she finally married him.[105] Early in November, 1863, she left her husband at Bannack to visit her family in Iowa.[106] The first news she received from him after her arrival at her early home was that he had been hung by the Vigilance Committee of Montana.

The report that Crawford had told the story of his intimacy with the Indian woman was fabricated no doubt by Plummer himself, and circulated for the purpose of involving Crawford in a quarrel in which he [Plummer] might kill him, without the odium attached of deliberate murder—and which might justify him too, more fully in the eyes of his affianced bride, Miss Bryan.

Crawford was wary, kept a sharp lookout, until on the 6th day of March he heard that Plummer was patrolling the street with a double barrel shot Gun between his cabin and Crawford's butcher shop, in which, latterly, no business had been done. Crawford posted himself on the opposite side of the way, armed with a double barrelled rifle, and as Plummer came up the street, he fired the contents of one barrel of his rifle from the corner of Jim Harby's house, then used as a Restaurant. This shot took effect in Plummer's wrist, shattering the bone and nearly causing him to drop his shot gun. He wheeled in the direction of Crawford, who instantly fired a second shot but missed him, Plummer exclaiming at the same time "you cowardly son of a b——h." Plummer turned and

---

105. A cursory, apparently less well-informed account of the romance between Electa Bryan and Henry Plummer appears in Langford, *Vigilante Days and Ways*, 67-68. For a detailed firsthand source, including a full description of the marriage ceremony by a participant, see Thompson, "Reminiscences," *Massachusetts Magazine*, 6 (July 1913), 118-19; and ibid. (October 1913), 160-61.

106. According to Frank Thompson's account, based directly on his journal, the actual date of Electa Plummer's departure was September 2, 1863: Thompson, "Reminiscences," ibid. (July 1913), 123.

walked down the street towards his cabin; nor did he accelerate his pace a particle in leaving.

Crawford retreated back behind the buildings and in a moment entered the back door of my store, which was the fourth house from the scene of the attempted tragedy. In company with him was Frank [Ned?] Ray, who was attempting to moderate his Nervousness, which nearly overpowered him. He was pale as a sheet, said "he was not used to the Rifle," and "wondered how it was he had shot so wild." E. C. Stickney and Bill Goodrich were in the store at the time, and Stickney wanted him "to load up and go out and finish the job." I advised Crawford to get over his excitement before he took any further steps, for in the condition he then was, he could not have hit the side of a house, forty paces off.

If, after being wounded by a rifle bullet, to stand up squarely and without flinching receive a second fire from a person—your mortal enemy—standing less than forty steps from you: if this be courage or physical bravery, Henry Plummer possessed those elements of character in an eminent degree.

The circumstance created no small amount of excitement among the residents of the town, and so divided were the people in their sentiments in reference to it, that had Henry Plummer then resolved to lead an honest life, he would have found plenty of good men who would have clustered around, and sustained him by their warm and Constant friendship. So well had he succeeded in blinding the public as to his real character up to this time, that half the town thought he was "more sinned against than sinning," while the conduct of Crawford in this affair was commented upon unfavorably, as compared with the nerve and metal exhibited by his antagonist.

Plummer sent Crawford word the next day by Budd McAdow that within three weeks, whether his arm was well

Henry "Hank" Crawford, shown with his family in the 1880s, fled Bannack after Henry Plummer threatened to fight him with "Bowie Knives, or Pistols as he might determine." (Beaverhead County Museum Association, Dillon)

or not, he would meet and fight him with Bowie Knives, or Pistols as he might determine. Within a week or ten days Crawford, believing he would be killed if he remained in the place, of which I think there was no doubt, left for Fort Benton and down the River to his home in Wisconsin.[107]

Before Crawford left, E. C. Stickney, who had come from Salt Lake City with Jack Mendenhall, and was employed by him as an assistant in selling his goods, and who slept and messed with us, came into the store with Crawford, and commenced talking of the trouble between him and Plummer. Stickney, who was much excited, began to tell what he would do if he was in Hank Crawford's place. I told him he had better remain quiet, and said to him privately there were a number of persons in the store, some of whom would probably report his expressions and·involve him in a difficulty with Plummer, or his friends. Stickney had evidently a great desire to distinguish himself in some way as a Fighting Man. It was his first experience in a new mining community, and he longed for that kind of respect born of fear, which constitutes a main element in the rowdy character everywhere.

He says, "I'm the best little man in the town. My name is Stickney, E. C., with a capital E and I ain't afraid to waltz up to anybody. Hank Crawford is my friend, and ought to go and kill Plummer in his bed." However correct or otherwise he might be, I did not like to have him talk in this manner in my store, for I knew well that if I permitted it, my place would become before long the scene of a shooting match with some of the hot-blooded rascals in town—and perhaps of Murder. Impressed with this belief I tried to persuade Stickney to remain silent.

---

107. Regarding Crawford's fight with Plummer, Purple's eye-witness account should be compared with the hearsay, perhaps embroidered, and certainly bowdlerized versions provided by both Dimsdale and Langford. Dimsdale, *Vigilantes of Montana*, 39-46; and Langford, *Vigilante Days and Ways*, 148-57. See also Mather and Boswell, *Hanging the Sheriff*, 34-35.

In a few moments he, Crawford and myself were left alone, when the door opened and in came George Carhart, who either was or appeared to be partly intoxicated. Carhart was as wicked a man when drunk as ever lived in the West. He had come for the purpose of quarrelling with Crawford, but Stickney and myself done all we could to keep him quiet, and in a few moments I managed to have Crawford leave the place. George was well fixed for a fight, having thrown off the large U.S. Dragoon Overcoat he usually wore, and put on a loose black coat which buttoned over the large Dragoon Six shooter he had slung in front of his person.

In a minute he and Stickney were discussing the relative merits of their two friends, Plummer and Crawford. I believe Stickney thought he was not armed, so effectively did his coat conceal the Pistol, till he jumped from the bench on which he had been sitting, drew and cocked his weapon. Simultaneously Stick and I were at his side. Capital "E" jumped round on the earth floor lively and brisk, with at least 7/8ths of the desire for glory he had an hour ago, knocked out of him by the palpable facts of this new scene, in which he was a busy and principal actor.

The click of a Pistol when being cocked close to your person, and in the hands of a man you know means mischief, either sends the blood to your heart with a paralyzing effect, or flowing at an accelerated rate through your veins adds much to the celerity of your ordinary habit of motion. I had known the man at Gold Creek. His reputation was proverbial. Without a scruple, he would just as lief shoot us in the back as any other way. He was in his usual drunken ugly mood, and E. C., who would by this time have written the E in his name in the smallest kind of an italic, and myself, went through all the forms of remonstrance, expostulation, and persuasion before we succeeded in getting him to put up his Pistol and leave the premises. Not a soul entered the store for three quarters of an hour, while we were engaged in this pleasant pastime of preventing the daylight from shining through our bodies, from the Pistol bullets of this drunken desperado. He was not angry at me, and always before and after treated me in a friendly manner. His cause of offense was with Stickney, and I determined that such a scene should not again occur in my place.

Immediately after Carhart left, Stickney went out to report to Crawford what had transpired after he left. He returned to the store about dusk in the evening, when I informed him there was to be no more messing and fussing about my place, and with some warmth, perhaps bitterness, reminded him that he was too "brash" in his loud talk and of too light Calibre to be a rival for Fighting honors about Bannack City just at that time. Without intending or expecting trouble from him, presuming as I did [that] he had enough for the day, I was much surprised when he said that he was bound to take a turn with me, and whipping out a Deringer he had in his pocket, offered to "Waltz all around me." I caught up my Pistol laying behind the Counter where I stood, when he started for the door; and raising up his arm to fire as he passed out, a person just coming in seized it, upon which he withdrew hastily, I seeing him no more till morning. He then came to me [and] said he was sorry for making a fool of himself, insisted upon presenting me with his Deringer, saying that he was satisfied that what I had often told him was correct, "that he was not a fit person to carry arms about." I never knew him after to mix himself up in other people's difficulties, although the year following at Summit on Alder Gulch he shot a fellow who rushed into his

---

108. Stickney's career in Bannack City, including his discussion with George Carhart, has received no mention in earlier accounts. Likewise, his encounter with the Virginia City vigilantes has gone unrecorded except by Ed Purple.

boarding house and attacked him while in bed with his wife and child. For this the Vigilance Committee ordered him out of the Country, but their acting in this respect was harsh and severe, and I believe they became convinced of it and revoked the order.[108]

On the 7th of March, 1863, I purchased in company with H. F. Morrell one eight (1/8) interest in the Dakota Discovery Quartz Lode, paying to L. Johnson and Joshua Laffin Three Thousand Dollars therefor in good clean Gold Dust at the rate of Eighteen dollars per ounce.

On the 22d day of March, in Company with the same gentlemen, I purchased from Josiah Chandler another interest in the same Quartz Lode, consisting of one twenty-fourth (1/24) part, for which we paid in the same manner the sum of Thirteen Hundred & Thirty-Three 33/100 Dollars, or at the rate of $32,000 for the entire claim. I believed this was, if not the best, one of the best Quartz claims I had ever seen. The yield from the Dacohtah Discovery, from March 8th to Oct. 3d, 1863, was $7,054.36 in Gold, over and above all the expenses of labour, tools &c. in working it. The work during this time was not constant, and [during] April, May and June there was nothing whatever done on it.

Artist and miner Peter Tofft drew this view of Deer Lodge City on Cottonwood Creek, probably in fall 1866, about three years after Ed Purple last saw it. (Museum of Fine Arts, Boston)

CHAPTER EIGHT

⚜

Bannack City and Deer Lodge
April 1863

On the 2d of April I left Bannack City to go over to Deer Lodge after a pair of cattle which I had been compelled to leave there the fall before, they having strayed from the herd. My travelling companion I expected to be Granville Stuart alone; but as we mounted our animals, we were joined by George Carhart, who said he had business at Gold Creek, and would go along with us. We reached Big Hole about noon, and stopped to eat our grub, which as is customary in that country, a man carries with him, or else he goes hungry. Jim Minesinger and Fred Burr were getting out timber here to build the Toll bridge which subsequently spanned the river at this point. Saddling up our Horses preparatory to pursuing our journey after dinner, Wm. Graves, alias Whiskey Bill, a stranger to Stuart and myself, rode up and speaking to Carhart with whom he seemed to be well acquainted, said he would ride along with us, over to the Divide, as it was on his way to a place where he had left some of his friends at work prospecting for Gold. At night all four of us camped on one of the little water runs, of which there are several putting down from the Divide. Carhart and myself slept in my blankets, and Stuart shared with Graves a part of his. Graves, who had neither bedding or provisions with him, left us early in the morning after eating breakfast, starting off towards the head of Silver Bow creek to join his friends in that locality. Stuart, Carhart, and myself continued our journey to Johnny Grant's, at which point we separated from each other to go our several ways.

There was a deeper meaning in the apparently casual companionship of these two men, on this journey, to Granville Stuart and myself, than we were then aware of. The events that transpired in the Territory within the next eight months conclusively proved that Carhart and Whiskey Bill were at this time members of that band of Highway men & Robbers, then in an embryo state of organization, but which soon after spread terror and death over the whole country.

I had no use for money on this trip, and had but seven Dollars in Gold Dust in a buckskin purse on my person, a fact which I had taken pains to inform my companions of immediately upon leaving Bannack. Stuart was as poorly provided as myself, having less than twenty dollars. The motive for murder was wanting, and to this fact I ascribe our safety. Our sleeping together was a measure of security, as I had accustomed myself to the habit of waking instantly, upon the slightest disturbance about camp at night. Had Stuart or I had Five hundred Dollars with us that night, I do not doubt but what the attempt would have been made to rob and murder us. Whiskey Bill when he left went to the rendezvous of his robber companions, who had their camp in one of the mountain gorges near the sources of the Silver Bow, or what was then called the first crossing or main Stem of Deer Lodge Creek. I afterwards learned that Reeves, Moore, Forbes, and others of the gang were at this camp at this time.[109]

---

109. Granville Stuart states that this trip began on April 24. For his account, which agrees with Purple's narrative on most essential points, see Stuart, *Forty Years*, 1:234-37. Other authors make no mention of the episode.

I remained at Deer Lodge two or three days, looking up my cattle, only one of which I was able to find. The other I never saw again. Having made an arrangement with Louis Demorest, or Damar as he was called, one of the old French settlers near Johnny Grant's, to drive over my steer with a band of his Cattle he had sold to the butcher at Bannack, I started back.[110] Reaching Fred Burr's on the Big Hole late in the afternoon, I stopped for the night. Here I found Tom Pitt, who had come over to look after his stock, which Burr had been taking care of during the winter. My Poney was jaded by the rough riding I had given him for the last week, and I concluded to leave him in the care of Burr, who had also been taking care of some of my cattle, and who gave me one of his horses to ride to Bannack.

This was the last time I ever saw my noble little horse. In less than a week, a thieving party of Indians (Burr said they were Crows) made a raid upon the cattle and horses in Burr's charge, and stole my Poney among a number of other animals which they succeeded in driving safely off. He had been my faithful companion for nearly a year in my wanderings, was as tough a piece of Horse flesh as ever was wrapped in hide, kind and gentle as a dog, never afraid to cross a stream or follow the most difficult and rugged mountain trail, never wandering off from camp or bothering me to catch him up, but always conducting himself like a horse of good sense and spirit, as he was. I could ride him with a wrapping twine for a bridle, and he paid no more attention to a rifle or Pistol when fired from his back, than as though he was a block of stone. Before leaving Salt Lake City I rode him from that place to Fort

---

110. Louis Demers, a native of Quebec who arrived at Deer Lodge in 1860, subsequently had an active business career that took him from Chicago to Corinne, Utah, and finally to Butte, where he died in 1910. His ties with Purple no doubt included their common identity as members of the Masonic order. For biographical information see Leeson, *History of Montana*, 1331; Anaconda *Standard*, March 3, 1910.

Crittenden and back around by Utah Lake, up to Salt Lake City again, a distance of ninety miles, in twenty-four hours, which included the time of rest we both got on the trip; and the day after I rode him out of the City on my way to the Salmon River mines, reaching the Train, which had preceded us three days before, at Birch Creek, a distance of nearly 40 miles. It has always annoyed me to think that my poor Poney, who had done me so much service, should fall into the hands of the Indians, to be made in his old age a Squaw horse of, by the miserable savages who stole him, or perhaps reserved for a worse fate, to have his bones picked for food by the hungry Indians.

Tom Pitt and I rode over together from Big Hole and reached Bannack about four o'clock in the afternoon.[111] We found the people of the town labouring under the great excitement about the killing of "old Snag," an old Indian who had been murdered a few hours before by Hays Lyon and Buck Stinson. Three or four Indian Lodges had been pitched down the creek for several days before, with the inmates of which the most friendly relations had been cultivated by the whites in and about the town. About noon old Snag, in company with his son, came into the place to visit Jimmy Spense's squaw, who was reputed to be old Snag's daughter, and who lived in the rear of Phil's Butcher shop. His son left him here, and shortly after the old man started up to see a little Indian girl living with Mr. & Mrs. Caroll, whose house was in the rear of the town, at the foot of the hill, over which the main road runs to Rattlesnake.

As he dismounted from his horse, Buck Stinson and Hays Lyon stepped up to him and made the remark, "We will learn you to kill white men," one of them put his six shooter to his

---

111. The name Thomas D. Pitt is included in Purple and Sander's census listing of Bannack City residents in 1862–1863, but biographical information is lacking. *Contributions*, 1:340.

head and deliberately blew his brains out. Old Snag had a rifle encased in a buckskin cover in his hands, but on discovering their intention to murder him, as they approached he rested his gun on the ground, leaning against his body, and raising his hands said "Me good Indian. Don't do it. Don't." But the words were hardly out of his mouth before he lay gasping in death. At the same moment, and as if by preconcerted action, a cry was raised in different parts of the town, that the Indians were murdering the whites. Believing this, many of the best disposed citizens rushed into the street with their firearms, and seeing the few Indians running in every direction and scampering up the hill sides, commenced firing at them, only one Indian, a Brother-in-law of Johnny Grant, being wounded by these general and indiscriminate shots.

Old Snag's son, who at a short distance had witnessed the attack and fall of his father, jumped on his horse and started up the hill; but before reaching the top he wheeled, and galloped like lightning back to the spot where his Father lay, and satisfying himself that he was dead, remounted and rose swiftly to the top of the hill. Here he jumped from his horse and shook his blanket several times in defiance at the men who were firing at him, and then without evincing any trepidation, he wiped his hands—which were covered with his Father's blood—on his Poney's mane. He was thus exposed for five or six minutes to the firing of the crowd below, and it was estimated that at least 200 shots were discharged at him during this time. After wiping the blood from his hands he mounted his horse and galloped back into the hills.

During the excitement Cyrus Skinner went out and scalped "Old Snag," cutting off a piece of his scalp from the top of his head of half the size of the palm of your hand, which for a long time after he kept hung up in his Saloon, as an ornament to his bar. After the excitement had subsided, the general opinion among the better class was that this was a brutal and cowardly murder—but as the victim was an Indian, the prevailing prejudice against the race stifled the voice of justice.[112]

Tom Pitt was very indignant at the occurrence, and it was no doubt well for him and me that the number of Indians who had been driven from the town were small, as had they been large enough to have scattered about the country, some of them might have fell in with us, on our way from Big Hole that day, in which case Pitt and I would probably been the victims of their retaliatory vengeance.

A day or two after, George Carhart returned to town and expressed himself freely in regard to the transaction. He denounced Stinson & Lyon in the most blasphemous terms for their cowardice, and it was supposed a fight between these worthies would be the Consequence. But Plummer finally fixed the matter up, and Carhart ceased to talk further about it, himself shortly following Old Snag to "that bourne from whence no travellers return."

Early in the month of April, 1863, a party consisting of Lem Nuckols, Jo Carrigan or Buffalo Jo, Bill Mitchell, Tom McFallen, John Brown, Buck Smith, Dick, a half-breed Deleware or Cherokee Indian, and three others left Bannack for the purpose of going to the Boise Country. They started out with the intention of going across the Mountains from Fort Limhi to West Bannack near Fort Boise. Jo Carrigan had kept a kind of sporting hall in the town during the winter, where he and Tom Foster had given lessons to such of the "natives" as desired to become proficient in the exercise of their "maulers," and perfect themselves in the manly art of self-defence. A few weeks before leaving, Jo from his supposed prowess at a rough and tumble fight, had incurred the ani-

---

112. Less detailed descriptions of this episode, generally consistent with Purple's hearsay version, appear in Dimsdale, *Vigilantes of Montana*, 51-52; and Langford, *Vigilante Days and Ways*, 178-80.

mosity, or jealousy, or both, of George Ives, and who disliked the "shoulder hitting" style of which Jo was at that time par excellence the professor and exponent in Bannack City. The two men met in Skinner's saloon, had some words, and Jo proposed to fight it out by a "square stand-up fight." But George had no idea of meeting the Champion fisty-cuff in this manner, and announced his purpose of killing Jo whenever or wherever he should thereafter meet him. On getting away out of the saloon Jo kept himself close in his Cabin for a number of days, and finally determined to leave a place where his life was constantly in danger of being sacrificed.

After the party had been gone two or three days, Lem Nuckols returned; the others kept on their journey. This was the last that was ever heard from them. In June a report reached Bannack that the skulls and other bones of a number of men had been found in the mountains, laying off in the direction they had travelled, and it was supposed that they had been met and massacred by a party of Indians in retaliation for the murder of members of their tribe by Moore & Reeves during the winter.[113]

The party were poor, had but little money with them, and their outfit offered little inducement to the merciless cupidity of the thieves and murderers who had begun to flow into the country. Their melancholy fate had been prophesied by many who deemed the time an inpropitious one for such a journey, owing to the grave causes the Indians had for active hostilities toward the whites, and for which the latter were justly responsible at least in the vicinity of Bannack.

In April a hundred or more Indians with their Lodges and families came to Bannack and encamped on the Creek above the town. Old Winnemuc, a Ute [Paiute] chief, was in the

---

113. The Purple and Sanders census listing identifies Joseph Brown, Joe Carrigan, William Mitchell, and "Smith, _____" with the notation "Killed by Indians, on Salmon river, in March, 1863." *Contributions*, 1:335-41.

party, and seemed to be the only noted Indian among them. This was the chief who had led the Indians at the battle of Pyramid Lake in Nevada, where Col. Ormsby had commanded the California and Nevada Men, and in which young Henry Meredith of Marysville, Cal., had lost his life.[114]

There was a general bitterness of feeling among people who lived in California at that time, against Old Winnemuc, and exhibited by none in a greater degree than by the rough element from the "other side" then living in Bannack. The Indians came into town to trade, and with them old Winnemuc, who had a contemptuous opinion of the whites generally, and a very high one of his own powers and dignity. It was said that he laboured to impress his warriors with the belief that he was invulnerable to leaden bullets, that he was a brother of the sun, and under the particular protection of the Great Spirit, that his Medicine was a great deal stronger and better than anybody's else, owing to the powerful agencies he could call around him.

He and his Indians were quiet and peaceably disposed while in town, but some twenty or thirty more of the roughs determined to try the potency of Old Winnemuc's Medicine charms, and attack his camp at night and if possible massacre his entire party. Their plans however leaked out, through persons who were kindly disposed towards the Indians, which coming to the ears of the roughs, among whom was George

---

114. Purple refers to a fight along the Truckee River near Pyramid Lake on May 12, 1860, in which a Paiute force headed by war chief Numaga led 105 white volunteers into an ambush, killing 76 and wounding many of the rest. Henry Meredith was a popular young attorney at Virginia City, Nevada, and the law partner of William Stewart; his death was mourned by a legion of friends. In a second battle on May 31, a combined white force of 750 volunteers and regular army troops killed some 160 Paiutes with a loss of only two attackers. On the Pyramid Lake War of 1860 see Russell R. Elliott, *History of Nevada* (Lincoln: University of Nebraska Press, 1973), 92-94. See also Ferol Egan, *Sand in a Whirlwind: The Paiute Indian War of 1860* (Garden City, N.Y.: Doubleday & Co., 1972).

Carhart, Moore, Reeves, Skinner, Plummer, Stinson, and others, they prudently abandoned their design. The next day old Winnemuc and his Indians left. It was reported that they lay all night in the sage brush, intending to ambush the party if attacked, in which event they would undoubtably have got the best of the fight.[115]

It was supposed by many that this party of Indians were [soon after] the authors of the Massacre of Buffalo Jo and his men, they having left only a short time before for Boise, and probably meeting the Indians who were exasperated by their treatment at Bannack, they had fallen upon and murdered them.

Shortly after Buffalo Jo's party had started for Boise, Jim Geary, Jim Stuart, Bosdick, Riley, Geo. Ives, Barney Blake and nine others left Bannack, to prospect for Gold on the Big Horn, and other creeks in the Yellowstone Country. Their intention was to thoroughly prospect the tributaries of the Big Horn, and if not prevented by the Indians to extend their search for diggins into the Wind River mountains. What little was known of these regions was derived from the gossiping reports of old Mountain men, and vague stories of the Indians, which represented them to be rich in the yellow metal.

The principal men in this party were old miners who had worked for years in California and in Colorado since 1859–60, and during the past winter had been industriously employed in the Diggins around Bannack. Geary worked on the bluff just below the town, and Bosdick on Dacotah Lode No. 8, of which he was part owner. I supplied Bosdick with the principal part of his outfit, except his animals, and he agreed

115. Other, briefer accounts of the visit of Winnemucca and his people to Bannack City appear in Dimsdale, *Vigilantes of Montana*, 50-51; and Langford, *Vigilante Days and Ways*, 178-80. Langford's account merges the murder of Old Snag, which occurred a few days earlier according to both Purple and Dimsdale, with this abortive plan to attack Winnemucca's Paiute encampment.

if he found diggins to stake claims for Tisdale and myself, in consideration of my outfitting him. He was furnished with two month's provisions. It was the intention of the party to remain out at least that length of time, if they did not make a discovery of Gold in paying quantities before.

Upon their return they reported that about three weeks after leaving Bannack, they met a small party of Crow Indians, who told them they must not come into their country "to dig holes in the ground," that it was a bad thing for the Indians. They said "a few White men come dig holes, by 'n by heap white men come, take all-same as white man take Bannack and Snake Country; no good for Indian." The party were disposed to conciliate them, and seemingly acquiesed in their wishes, and the Indians left. They were however unwilling to abandon the object of their expedition, and Riley, who had been elected Captain of the Company, submitted the question to the men, who all resolved to go on prospecting as before for Gold.

Two or three days after, while in Camp on a gulch, a small tributary of the Big Horn, they were attacked by a party of Crows. The attack was made at night, the Indians firing into the Camp simultaneously with the alarm raised by the guard. John Geary and Bosdick were mortally wounded while laying in their blankets.

The attack was sudden and brief, only lasting a moment, but the savages were lurking about them in large numbers, ready to resume it, the next day. When daylight appeared, the party removed their wounded and dying Companions a short distance away, where they were less exposed to the fire of the Indians. They could not think of leaving them, and as their condition precluded their being removed from the place, they resolved to remain and protect them. Against this arrangement Geary and Bosdick urgently remonstrated. They said they knew they had received their death wounds, that at the best they could live but a few hours, and that it would be a certain but useless sacrifice of the lives of the whole party for them to remain on their account. They therefore implored that their Pistols should be handed them, and they [be] permitted to end their own sufferings, and thus relieve the party from the death in which they all would be involved, if they remained.

The Indians were around them, and hemmed in by danger on every side, the party felt that their only hope was in Complying with this request, to which they finally and reluctantly yielded. Geary blew his brains out at the first fire of his pistol. Bosdick placed his to his head, pulled the trigger, and the hammer failed to burst the cap. He looked up, smiled faintly, but recocking the Pistol immediately, discharged its contents through his head. Thus died two men whose heroic self-sacrifice, under circumstances so distressing, entitle them to the venerating regard of the first settlers of Montana.

The balance of the party, though sorely pressed by the Indians for a week after this occurrence, finally reached Bannack without any additional loss to their numbers.[116]

While this party was away, another left Bannack and started towards the Madison River country for the purpose of prospecting for New Diggins in that region. It consisted of six or seven persons, among whom was Wm. Fairweather, Thomas Cover, [Henry] Edgar & Barney Hughes.[117] I was not personally acquainted with any of them except Fairweather and Edgar. How far east of the Madison River, into the Yellow Stone Country they went I do not know, but their search for Gold in that direction was not successful, and they started

---

116. For the principal primary account of the Yellowstone Expedition of 1863 see James Stuart's journal, *Contributions*, 1:132-205.

117. In his journal of this expedition, Henry Edgar supplies also the names of Henry Rodgers and Bill Sweeney: "Journal of Henry Edgar—1863," *Contributions*, 3:125. The account by Peter Ronan identifies the same six men as members of the discovery party. "Discovery of Alder Gulch," *Contributions*, 3:143.

William Fairweather discovered the gold-bearing rimrock that spurred the rush to Alder Gulch in May 1863. (Thomas J. Dimsdale, The Vigilantes of Montana, 1915, opp. p. 127)

back. Their return was hastened by the hostile disposition of the Indians of which they saw plenty of sign, and one small party of whom they met.[118] One night in the latter part of May, 1863, they camped on the creek about a mile and a half above the hillside which is now the site of Virginia City. The next morning after breakfast Fairweather started down the creek to look up his animals, which had wandered in that direction and were feeding on the grassy slope of the hills, between which there was a small run of pure water, and upon which a city of 10,000 inhabitants was destined to be built up within 6 months by the Discovery of Gold he was that day to make. Finding his animals he started back for his camp, going

---

118. In fact, a band of Crow Indians took the Fairweather party captive and held them a few days, then released the gold-seekers after making them promise to leave Crow territory. Henry Edgar, "Journal of Henry Edgar—1863," *Contributions*, 3:124-42.

Henry Edgar, who posed for this photograph in Alder Gulch on August 29, 1899, was one of the discovery party at the new diggings, which they originally dubbed Stinking Water. (Montana Historical Society Photograph Archives)

up the left side of the creek, and when about a quarter mile above the point where the City now stands, he was attracted by the appearance of the rim rock that rose out of the ground on the opposite bank.

"There," says Fairweather, who related to me the circumstances of discovery while standing upon the ground two months after, "there, says he to himself, if there is any Gold in this section of the Country, it is right over there on the pitch of that rim rock." He went to his camp, got his Pick, Shovel and Pan, returned to the place where the "rim rock struck up," sunk a hole two feet deep in the ground, took a pan full and washed out over Four Dollars ($4) the first time trying. Several other pans were washed, varying from $2.50 to $6 per pan, showing him conclusively that he had made the discovery of uncommonly rich Gold Diggins. The party located their claims as Discoverers, and being in the want of provisions started for Bannack the next day to obtain more, and to report to their friends the rich discovery they had made.

Gold miners as a class are in no way credulous of reports which they hear about the discovery of new diggins. They believe in them implicitly, and the more marvellous the account is of their extent and richness, the more easily are they convinced of the truth of the story. The day after Fairweather & his party reached Bannack City, the report they brought had spread like wild fire among the Miners, and it lost nothing of its original brilliancy as it circulated from mouth to mouth. Good mining claims about Bannack that could not have been bought for less than $500 to $800 a week before went begging for buyers at $100 to $200, and in less than two weeks three-fourths of the population of the

---

119. "Journal of Henry Edgar," *Contributions*, 3:124-42. For the stampede to Alder Gulch, see Stuart, *Forty Years*, 1:246-47; Dimsdale, *Vigilantes of Montana*, 71-74; Langford, *Vigilante Days and Ways*, 206-7; and Ronan, "Discovery of Alder Gulch," *Contributions*, 3:148-51.

town had stampeded for the new mines on the Stinking Water.[119]

Fairweather and his party remained in town for two or three days, making their purchases of tools, provisions, &c., during which time large numbers of the Bannackites were making their preparations to accompany them. These were the busiest and liveliest times I ever saw in the town. The few goods were rapidly bought up by the Miners. Pack and riding saddles, Horses and Mules were in great demand and brought extravagant prices. Dick Hamilton, who had come in on the 30th of April with a stock of goods belonging to W. A. Carter of Fort Bridger, was kept up half the night, with John Sharp and myself as Assistants in his store, waiting on the men who flocked to it, to purchase their supplies. The fluctuation of the Gold Market in New York was nothing compared to the fluctuation of prices for merchandize in Bannack at this time.

One morning Dick Hamilton commenced selling Rope for packing at $1 per pound. At 10 o'clock the price was advanced to $1.50, by noon it was $2.00, and by 4 o'clock the demand was so great that a further advance was put upon it of 50 cents per pound, and continuing to sell rapidly, Dick put up the price to $3.00 per pound, at which rate it continued till the close of business at night; when the next day it was offered and sold out at $3.50 per pound.

Upon the third day after their arrival Fairweather & party started back, taking with them at least two hundred people, to the new mines on Alder Gulch. Wm. Fairweather, the first discoverer of these mines, secured for himself a first rate claim which he worked for more than a year, and out of which he made a large sum of money. He did not, however, save but little of it from the human leeches that fastened upon him, when his good fortune had made him one of the wealthiest Miners on the Gulch. He sold out his interest in his mines in the fall of 1864, and was soon after reduced to the same pecuniary condition he was in when he worked for daily wages [at] Gold Creek two years before.[120]

In April I rented my store to Robt. Hereford, who came in this spring from Fort Bridger. He had a fine Indian Lodge made of dressed Buffalo Skins, which was pitched back of the store, and under which Tisdale, Sam Batchelder and I slept for six or seven weeks.[121]

One night we were awakened by the firing of a number of shots in the street, and in a few minutes we heard the voices and footsteps of several persons approaching the Lodge. I called out "who is there," when Henry Plummer with a lantern in his hand, threw the Flap of the Lodge to one side, and thrusting in his head, wanted to know if we had seen or heard any one prowling about there within a few minutes. Telling him we had not, he said, "there has been a fight over in Skinner's Saloon, and George Carhart has been shot, and I am looking for the son of a b——h that done it." With him was Charley Reeves, Moore, and Forbes. That night we had with us in the Lodge an elderly man by the name of Ettrick, who had just come into Bannack, and [he] being flat broke I had hired him to get timber for my new house. As Plummer turned away from the Lodge, Ettrick hollered after him "will the fellow die?" There was in the tone of Ettrick's voice something which Plummer interpreted as unfriendly towards Carhart, and stepping back to the Lodge, with curses such as none but the gang of men he belonged to could spit from their mouths, inquired "which son [of] a b——h that was that

---

120. Addicted to alcohol and his fortune gone, William Fairweather ended his days depending on the charity of old friends in Montana. He died at Passaman valley and was buried in Virginia City in 1875. A death notice appeared in the (Virginia City) *Montanian*, on June 25, 1875; the same paper published an obituary on September 2, 1875.

121. Robert Hereford arrived in the upper Bitterroot valley in 1856. He later became Helena town constable and Lewis and Clark County Assessor. On his career see Judge Frank H. Woody, "A Sketch of the Early History of Western Montana," *Contributions*, 2:95.

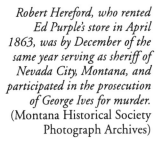

Robert Hereford, who rented Ed Purple's store in April 1863, was by December of the same year serving as sheriff of Nevada City, Montana, and participated in the prosecution of George Ives for murder. (Montana Historical Society Photograph Archives)

spoke." Ettrick covered his head in his blankets and remained silent, while we pacified Plummer by telling him that the man was drunk, and that he must not take any notice of him.

In the morning we learned that a party, consisting of Richard Sapp, Banfield, Bissell, & Reeves, had been playing a four-handed game of Poker in Skinner's Saloon. A dispute had arisen between Dick Sapp and Banfield in regard to the play, Sapp accusing Banfield of cheating, when they jumped from their seats and fired all the shots in their revolvers at each other, but without effect. George Carhart was lying on a table in the front part of the saloon, drunk and asleep. It was soon discovered that he had been shot by a bullet from one of the pistols, and Moore, Forbes, and Reeves immediately pulled out their Six Shooters and commenced firing at Sapp and Banfield to revenge his death.

Moore told me that he fired one shot at Sapp and missed him, on account of the dull light. The next wounded him slightly in the hand. Moore then stepped up to him, and held the Muzzle of his Pistol close to his face. Says Moore, "he never batted his eye, and I did not have the heart to kill as brave a man as he appeared to be." He lowered his pistol and let Sapp walk away out of the Saloon. In the meantime Banfield had rushed to the door, and escaping the flying bullets in the saloon, was struck by one just as he stepped out which shattered his leg at the ankle joint. He managed to crawl in behind the buildings and hide for a time, and finally dragged himself down to a cabin in the lower part of the town, which he reached without being observed by the party in pursuit of him.

During the melee a man by the name of Reilly, a looker-in, was wounded in the leg but not mortally. Carhart was taken across the way and died within a couple of hours, lamenting that "he had no show for his life and must die like a dog."[122]

After Banfield had laid for a week, Dr. Glick, Leavitt & Young came to the conclusion that to save his life, his leg must be amputated. This was for some reason deferred so long that gangrene had set in, and extended to the body, so that all hope was cut off. He lived about twenty days after he was shot, Cursing nearly all the time when awake, and swearing that "he would not die till he had had satisfaction out of the sons of b—hs that shot him." He would doubtless have made a capital recruit for the gang of roughs had he lived.

It was said George Carhart had represented one of the Northern Counties of California in the Legislature of that state in the winter of 1852–3. He was a fine looking man, and when sober, quiet and gentlemanly in his behaviour; but whiskey proved his ruin. At the time of his death he was owner with George [Ives?] of Dacotah No. 7, one of the best Gold Quartz Lodes about Bannack. His friend Plummer

---

122. Other accounts of this story offer significantly different details. See Dimsdale, *Vigilantes of Montana*, 49-50; and Langford, *Vigilante Days and Ways*, 176-78.

123. For remarkably similar appraisals of George Carhart, see Dimsdale, *Vigilantes of Montana*, 49; and Langford, *Vigilante Days and Ways*, 178.

administered upon his estate, which however only consisted of this Lode.[123]

He reported before his death that he robbed Davenport over [at] Black Creek a few weeks before, and there was some at Bannack who thought that his death was not so accidental as reported, but was the work of Moore, Reeves and Forbes, because he had the habit of talking too much and letting out the secrets of the gang when drunk. From this fact it was supposed he was put out of the way. It did not do for a man to ventilate such subjects much, about Bannack in these days, and after his death & funeral but little attention was paid to the matter.[124]

In April E. C. Stickney was married to Miss Mary Donnelly, who was during the winter one of the belles of Bannack, the other being Miss Matilda Dalton. Up to this time the imps of darkness were ahead of all the evangelical Ministers in getting into the country, and no clergyman could be found to perform the marriage ceremony. Stickney was anxious—as many a man before and since his time has been—that the matrimonial knot should be tied at once; but in what way or by whom was a question which involved considerable difficulty in its solution. To aid him in his cherished and laudable design, and with feelings of the most sincere respect for himself and the affianced bride, I proposed to unite them in the holy bonds of wedlock myself, by reading to them the beautiful service of the Episcopal Church in such cases, made and provided for in the Common Prayer Book, of which I fortunately had a copy. Stickney had made up his mind to accept my proffered services, when J. M. Castner, the brother-in-law of his intended, suggested N. P. Langford, who had acted as Chaplain in Cap't. Fisk's train while on its way from Minnesota, as perhaps a more suitable

124. The suspicion that Carhart's death may have been more intentional than accidental is not reported in any other primary accounts of these events.

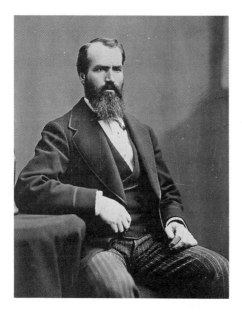

*Nathaniel P. Langford,
shown here in about 1875,
became a vigilante in the
heydays of Bannack and
Virginia City, then wrote
about it all in* Vigilante
Days and Ways, *published
in 1890.* (Montana
Historical Society
Photograph Archives)

person to bind them together in the enduring relationship of
husband and wife. Mr. Langford was therefore engaged for
the purpose. After the Ceremony an elegant supper, prepared
by Mr. and Mrs. J. M. Castner, was partaken of, and dancing
Commenced which was kept up to a late hour. The guests
were numerous, nearly everybody in town having been
invited, and the occasion will be long remembered as the
most elegant and recherche of a social character, ever before
had in Bannack, as well as for its being the first wedding ever
celebrated in Montana Territory.[125]

As the spring advanced, other rough characters came with
the first emigration that began to flow into the country, some
of whom had wintered at Salt Lake City. The sports and
pastimes of these gentry was gambling, strap and safe games,

---

125. While it is hardly accurate to say no weddings had been celebrated in the
Montana region prior to this event in April 1863, Purple evidently meant this
claim to apply to marriage ceremonies between non-Indian newcomer couples,
following the forms customary in the eastern states. Considering his claim, it is
all the more remarkable that the presiding officer, Nathaniel P. Langford, makes
no mention of the affair in his own chronicle of the early days.

and other kindred modes of fleecing the unwary out of their hard earned savings. Davy Morgan, an industrious and honest miner, was "beat" out of $150 by being too confident of Charley Forbe's honesty, and loaning him that amount to bet that the "relic of a dead Father or Mother" was not in the little safe which Davy thought had been thoroughly inspected by him and which he believed to be empty, having seen the "relic" fall to the ground from the safe Box a moment before.

Charley Forbes, alias Ed Richardson, was a man about 24 years of age, the superior of Reeves in education, had pursued his career of crime in California and Nevada a number of years, and was in all respects as accomplished a villain and cutthroat as ever escaped death by strangulation. He used his pistol in a handier manner than even Plummer, and it was said of him that he could empty his Six shooter a second or two Sooner than any man in Bannack City. He took his place at once, upon his arrival, as a star—if not the first, then one of the most brilliant of the second magnitude—in the galaxy of roughs. He was more temperate than any of his companions, and was never known to be drunk. He was a well informed man, wrote good letters, some of which were published in the Sacramento Union, a paper which came to him quite regularly for a couple of months after his arrival. He came to me one day and asked a loan of $60, stating that he had nothing but his honor as a sporting man (his ostensible occupation) to pledge for its return the next day at 4 o'clock. I gave it to him, but he did not return it till the morning after he promised, stating that he was busy in a Poker game at the time he should have returned it.[126] Ives, Wm. Hunter, and Carhart were different men, and owed me for provisions when they died.

---

126. For additional biographical information on Charley Forbes, an alias for Edward Richardson, and on Forbes's death at the hands of fellow gang member Gus Moore, see Langford, *Vigilante Days and Ways*, 214-16, 219.

A vigilante execution (Nathaniel P. Langford, *Vigilante Days and Ways: The Pioneers of the Rockies*, vol. 1, 1890, opp. title page)

CHAPTER NINE

⚜

Bannack City
May to September, 1863

In the month of May Plummer, who had recovered as much as it ever was possible for him to do from the wound in his wrist, under the surgical advice and treatment of Dr. Glick, inaugurated a movement by which he was made the sheriff of Bannack City. In March Congress had detached that portion of Dakota [Territory] in which we lived, and in the organization of Idaho, Bannack City came within the boundaries of that Territory. Up to the present time the people here had lived without the forms of law, save in respect to mining property, and it was felt by all that there was great necessity for the establishment of judicial tribunals, and a general organization of those safeguards of life and property which are usually relied upon in all civilized communities for the preservation of public peace and order. A meeting of the citizens was held in May, at which there was probably 150 people who briefly resolved that an election of a Judge and sheriff should take place in three day's time. The Judge was to have jurisdiction in all cases of felony, and such actions to determine the right to Mining claims as parties might choose to commence in his court. B. B. Burchett immediately announced himself a candidate for Judge, and Plummer and Jack Sheen or Shaw became rival Candidates for the office of sheriff.

The day before the election came off, Plummer came to me and urged me strongly to run as Judge, having learned that I acted as Justice of the Peace in Calaveras County, Cal. I declined, but he pressed the matter so far, that he insisted upon posting my name as a Candidate, and declaring that I would surely be elected. I told him that whether elected or not, I should not serve in such Capacity. I suppose he was desirous to make as much excitement over the election as possible, but at the same time secure his own and Burchett's position as Sheriff and Judge. Some of my friends, thinking I wanted the office, rushed to the polls, and despite my protestations that I would not serve, came within twenty or thirty votes of electing me.

Plummer was elected by a round majority, there having been over three hundred votes polled.[127] After the election Plummer, who since his shooting scrape with Crawford had demeaned himself with the utmost propriety, started for Sun River, married Miss Bryan, her brother-in-law officiating as Clergyman, and returned to Bannack, where he was well received by the people, and at once assumed the duties of his office.[128]

Among his appointments was that of Wm. Dillingham, from the other side, as Under-Sheriff. I believe Dillingham

---

127. The Bannack Mining District Records, SC 238, MHS Archives, Helena, contain the proceedings of a meeting at Bannack City on May 24, 1863, held "for the purpose of electing a Judiciary and Executive for Bannack District." At this meeting, chaired by district president W. B. Dance, the assembled residents resolved that the offices of judge, sheriff, and coroner should be elected by ballot. Dance then appointed Ed Purple and two others as election judges, and also appointed three tellers. Voting lasted two hours, with 554 votes being cast. At the end of that time, the tellers reported that B. B. Burchett had been elected judge, H. Plummer elected sheriff, and J. M. Castner elected coroner. The assembly then resolved to declare these three men elected unanimously.

128.   Frank Thompson's account confirms that Father Joseph Menatre officiated at the Bryan-Plummer wedding ceremony. Thompson, "Reminiscences," *Massachusetts Magazine*, 6 (July 1913), 118–19.

Walter B. Dance appointed Ed Purple and two other men as election judges for the balloting that made Henry Plummer sheriff of Bannack City. (Nathaniel P. Langford, *Vigilante Days and Ways: The Pioneers of the Rockies*, vol. 1, 1890, opp. p. 308)

undertook the duties of his office with the honest intention of discharging them for the good of the people at Bannack. I believe that he, in common with the great majority of the citizens, was ignorant of Plummer's true character, and thought he would govern himself in his office with a sincere desire to detect, prevent and punish lawlessness and crime. I do not wish to burden the name of a dead man by charging upon him a crime of which he may have been guiltless. A few more or less would have in no way relieved the blackness of the Sins that stained the soul of Henry Plummer. He undoubtedly instigated the murder and death of Dillingham. In appointing him as a Deputy, he mistook the virtue and honesty of the man. Dillingham, with a rectitude of purpose which will ever commend him to those who knew him, set about diligently the performance of his duties as Conservator of the public order, and as far as it lay in his power, the suppressor of crime in its various forms in the Community. Plummer soon found that it would be dangerous to attempt to mould him to his purpose, and without ever confiding to

him the secret of that organization for Highway Robbery and
Murder, then in the incipient stages of its formation—and of
which he was the acknowledged leader—determined to rectify
the mistake he had made and consign Dillingham to a bloody
grave.

Meeting Dillingham one day in June, he says to me "there
is a lot of Robbers about this town, and I intend to nip some
of them one of these days. I have overheard them, talking over
their plans, and am going after them." I have no doubt that
this language, communicated by him to his chief, the High
Sheriff of the County, Henry Plummer, the man of all others
in the County who should have put forth every effort to aid
and assist his subordinate, sealed his doom.

The day after this conversation, Dillingham started for
the Stinking Water mines, discovered a few weeks before and
called afterwards Virginia City. Before leaving I heard him say
that he was going to arrest some thieves or highway robbers.
Up to that time I had heard of no Road robberies, but one or
two cabins in Bannack had been broken into by thieves, and
Mrs. LeGrau had had some jewelry and a small amount of
Gold Dust stolen.

This discovery of rich diggins on the Stinking Water [at
Alder Gulch] had drawn off more than half the population of
Bannack City, and with it a large number of the gang (by
which name I shall hereafter designate the Gamblers, robbers
and Murderers of whom Plummer was chief), who now
began to divide their time between the two places, and points
along the road leading to and from them.

It was about this time Hays Lyon became a member of
the gang. He had driven an ox team for Jack Mendenhall into
the country the fall before, and during the winter had worked
industriously at Mining. He was a light haired, fair complexion, rather small featured, and small sized man of 22 or 3
years of age. When he first came into the country he was pure

On January 14, 1864, when the building in this picture was under construction, one of its beams served as the gallows for the hanging of five road agents, including Hays Lyon. (Montana Historical Society Photograph Archives)

and unsullied from crime, and had he not sought the companionship of the profligate and depraved, he might have lived as he first started out to live, an honest and respectable member of society.[129] "As there are heights of glory, so there are depths of infamy," and into the deep abyss in which his companions revelled young Lyon had not yet plunged. Another beside him, Buck Stinson, a recent addition to the gang, had not yet imbued his hands in the blood of a white man, and as neither of them in the vernacular of the gang, had had "a man for breakfast," they were selected as the proper ones to murder Dillingham.[130]

There was an additional reason for their being assigned this bloody work. The opportunity of showing themselves worthy of the consideration of the gang had not been afforded them, save on one instance—that of their murdering a month before an Indian named Old Snag, a petty chief of the Bannack or Snake tribe. His unresisting and passive action, when attacked in the streets of the town, his advanced age and defenseless condition, his appeal "me good Indian" while they strode upon him, and holding their pistols to his head, sent their bullets crashing through his brain—all this which strongly commended them for bloodthirstiness failed to satisfy the gang that they were possessors of the nerve and courage which ought to distinguish their fraternity. "It was a damn cowardly murder," said George Carhart, and in this opinion the citizens generally coincided. The people had however learned but a short time before of the fate of Buffalo Jo and his party, who had been killed by the Indians, and the murder of old Snag was tacitly approved as a retaliatory

129. For a brief biographical sketch of Hays Lyon, based on the Dimsdale and Langford accounts, see Thrapp, *Encyclopedia of Frontier Biography*, 2:888.

130. For a detailed biographical sketch of Buck Stinson see Thrapp, *Encyclopedia of Frontier Biography*, 3:1371. For one traveler's encounter with Stinson and his companions shortly before a squad of vigilantes overtook them, see Toponce, *Reminiscences*, 55-66.

measure, and which these worthies strongly urged as a defence for their inhuman and cowardly conduct.

The gang being divided somewhat in sentiment as to this deed, to better prove their valor and pluck with bloody hands new laurels to twine their murderous brows, Lyon & Stinson left Bannack and followed Dillingham to Stinking Water. With them went Charley Forbes, who was an old hand at such business, and who, if the other two should "weaken" or lack the "sand," would see that the victim did not escape the death intended for him. On their way over they reported that Dillingham had said they were thieves and Robbers, and the same report was circulated in Bannack, after their departure, by Plummer and their friends—for the purpose no doubt of excusing the murder they knew well was so soon to follow.

Upon reaching Stinking Water, they found Dillingham sitting in a brush house, on the Creek, near the part of the main street of what after became Virginia City; and calling him a few steps from the side opening or door, Lyon spoke hurriedly to him and said "damn you, you have been lying about us," fired at him slightly wounding him. At the same moment or instant of time Stinson exclaimed "Take back them lies you told," and also fired without much effect, when Charley Forbes finished the job by sending a bullet through his heart, and the first Under Sheriff of Bannack City lay dead at their feet. The first shot was fired while Lyon was speaking to him, and the others followed in such rapid succession that poor Dillingham had not even an instant of time to reply. It was not their intention that he should, for he might have said something which would have been heard by those within the Brush House, and the dying words of their victim might have irrevocably sealed their fates and disrupted and ground into dust that embryo organization of villainy and murder which scoured like a flaming sword for months the inhabitants of this new Territory.

Plummer had correctly weighed the chances of Lyon &
Stinson, mere novices in the art of Murder, killing
Dillingham, and there is little doubt, had they not been
accompanied by Forbes, from the wild and random way they
shot, they would have had more difficulty in disposing of him
than was the case with Forbes as helper and backer. It was in
fact him who done the deed.[131]

For this murder, Lyon, Stinson & Forbes were immediately arrested and put upon trial by the miners of Alder
Gulch. Forbes was acquitted and used to say after at Bannack,
"I talked too well for them to hang me." The other two were
convicted and sentenced to be hung. A gallows was erected on
the hillside a short distance from the creek, and the graves
dug to receive their bodies. On the way to the gallows only a
small portion of the crowd of miners who had been present at
the trial accompanied the culprits. A few women were in the
procession, which stopped a short distance from the intended
place of execution, and someone mounting the wagon in
which Lyon & Stinson were seated, commenced reading a
letter purporting to have been written by Hays Lyon to his
Mother, but which some friend of his was really the author.
The letter [was] affectionate in its terms, and asservating in
the strongest manner the innocency of the writer of the crime
for which he was about to die, brought tears to the eyes of the
females, and earnest pleadings from them and others that the
execution should be deferred. A vote was taken whether to
hang or let them go. The crowd was small, their friends and
sympathizers were in the majority, and they were set free.
They made hasty tracks from the place where they had come
so near finishing their earthly careers. Had the Miners in the
gulch been present in as large numbers as at the trial, they

---

131. Accounts of the murder of Dillingham, substantially in agreement with
Purple's narrative, appear in Dimsdale, *Vigilantes of Montana*, 74-76; and
Langford, *Vigilante Days and Ways*, 207–12.

would have been hung. Lyon afterward made use of the grave dug for him as a place to relieve the natural want of nature, a fact which he jokingly communicated to several of his friends. But he went not long unwhipt of justice, for in less than eight months his neck was stretched with a cord, and his body was laid in the identical grave originally designed for it, and which he had so often jocosely desecrated.[132]

I had my new house finished by the last of June, and rented it to A. J. Smith and Brown, a Coloradian, for $200 per month. It was 28 feet wide, 58 feet long, and 14 logs high. The logs had been hewn on two sides, and when completed the house was one of the best looking and had the largest single room in Bannack City. Brown had brought his Billiard Tables with him, and the place was used as a Billiard Saloon. The excitement attending the discovery of the mines at Stinking Water or Alder Gulch had by this time driven away nearly the whole population of Bannack, and the town was nearly deserted by the 15th of July. Smith & Brown occupied my house only for a month, after which it became like three-fourths of the others in Bannack, empty.

John Sharp, an old friend of mine who had come to Bannack on the 30th of April with R. H. Hamilton, in charge of a large assortment of goods, groceries, &c., belonging to W. A. Carter of Fort Bridger, wanted me to assist him in the store, which I done. I remained in this employ till I went to Virginia City, on my way home to New York the following December. Hamilton took a wagon load of goods over to Alder Gulch when Fairweather and his party first went back there, after their discovery, and this was the first loaded wagon that ever made a track into that Gulch.[133]

---

132. For the trial and release of Stinson, Lyons, and Forbes, see Langford, *Vigilante Days and Ways*, 212–18, and Dimsdale, *Vigilantes of Montana*, 76–82. On Lyon's grave, in more circumspect language, see Dimsdale, *Vigilantes of Montana*, 168.

In July the emigration up the Missouri River began to arrive, but most of the people went directly to the Stinking Water mines. A few came to Bannack, among whom was James Tufts, D. H. Hunkins, Dorson Chase, John Ely, Henry Ohle, Cyrus Gilbert, and Henry Thompson. From Salt Lake City and by the land route then had arrived M. J. McDonald, George Chrisman, Gov. J. Duane Doty, Charley Ohle, Frank Gilbert, John Bartelson, Mr. & Mrs. George Dye & family, Morgan, Mrs. Belt, widow of Dr. Belt of Weston, Mo., and her daughter Miss Nanny, who married Horace Wheat the next year. From California had come my old friends George Batchelder, T. W. Taliaferro, Dave Phillips, Solon Pratt, J. P. Haskell, and a number of others. A few of these remained at Bannack, but most of them located at Virginia City.[134]

In July & August Bannack, from the paucity of its population, was as quiet as a graveyard. Plummer had his headquarters here, and had succeeded in getting himself elected Sheriff of Virginia City. This was accomplished by the gang, and under the supposition that he would take up his residence there, but he still Continued to live at Bannack. He would absent himself for a day or two at a time, no doubt looking after the interests of the band, but at the time giving as an excuse that he was looking after a new discovery of some

133. Peter Ronan, who was present, declares that three days after the stampeders had staked their claims, James Sheehan drove the first wagon into Alder Gulch. This wagon carried Sheehan's wife and children, including a daughter who later became Mrs. Ronan. Ronan, "Discovery of Alder Gulch," *Contributions*, 3:151.

134. Historically, the most notable persons in this listing are James Tufts, later secretary and acting governor of Montana Territory, and James Duane Doty, former governor of Wisconsin Territory who had been appointed governor of Utah Territory in early June 1863. On the career of Governor Doty see Alice Elizabeth Smith, *James Duane Doty, Frontier Promoter* (Madison: State Historical Society of Wisconsin, 1954). Smith makes no mention of Doty's 1863 trip to Montana. Information is scarce on the other individuals listed here by Purple, although George Chrisman's brief reminiscences appear in Robert W. Richmond, "A Pioneer's Report—1865," unpublished manuscript, Chrisman Vertical File, MHS Library, Helena.

Quartz Lode, reports of which were in constant circulation, and thus his absence was regarded as an ordinary and commonplace circumstance. At this time there was never more than three or four of the gang ever seen together. Geo. Ives and Crow, alias Texas, had established themselves as herders on the Madison River, some 25 miles from Virginia City, where at so much a head per month they took horses and cattle to Ranch. Persons who left horses with them would inform them the day before when they wanted them for use, and as one or the other was in town every day, they could easily be brought in, which saved the Miners and others much trouble in looking up their own stock. Others of the gang were located in different places about the country, engaged in the same occupation or travelling about as miners, the roving habits of the people generally affording sufficient excuse for their actions without exciting suspicion of their real objects and designs.

Some were in the towns of Bannack and Virginia City all the time, engaged in gambling, and picking up such items of information as to when men with Money were leaving one place to go to another, and posting their companions out of town in regard to the movement of every party who was known to have Money, and who was about leaving the Country.

In September Ben Peabody's Coach, that ran from Bannack to Virginia, was stopped by two or three men, and the passengers robbed. Billy Rumsy was the driver, an honest little fellow, and if he discovered beneath the black neckerchiefs that were tied over their faces who the Robbers were, it would have cost him his life to have divulged the secret. An Irishman generally known as Bummer Dan was the principal sufferer by this robbery. He had a rich claim back from Alder Gulch, which he worked with great profit, and it was supposed he had a large amount of money with him at the time, but the amount was less than $2000. It was reported that

A visitor regards Henry Plummer's grave in Hangman's Gulch with Bannack in the background sometime after 1910. (E. C. Schoettner, Butte, photographer, Montana Historical Society Photograph Archives)

Bummer Dan said he recognized one of the robbers as George Ives, but in a day or two this merged into a mere rumor, which Dan, fearing for the consequences, pronounced as being untrue.[135] This robbery created more alarm than anything that had transpired in the country. Yet men spoke of it only in whispers and to their most intimate and confidential friends. From this time forward until the organization of the Vigilance Committee in December and the hanging of Ives, Murders and robberies on the highway succeeded each other

135. Purple's brief telling of this episode may be compared with that of Dimsdale, *Vigilantes of Montana*, 64-70; and Langford, *Vigilante Days and Ways*, 233-40. Dimsdale places this holdup in late October and supplies circumstantial detail. He claims that Bummer Dan—Daniel McFadden—along with one other passenger, knew the identity of the stage robbers (George Ives and Frank Parrish), and told it to a third passenger, but no one divulged the secret until after the two men had been hanged by the vigilantes. Langford's account follows the facts set down by Dimsdale, but adds more melodrama.

in frightful rapidity. Henry Plummer, the spirit who ruled and directed this band through all these scenes of bloodshed and robbery, was as tranquil and mild about the streets of Bannack, as though he was as innocent of their commission as the most worthy man in the Community.

⚜

Ed Purple's narrative ends at this point. The following fragment, included with his unfinished manuscript, serves in effect as addendum to Thomas J. Dimsdale's Vigilantes of Montana. *It provides further substantiation for the belief that Henry Plummer had been at the center of a criminal conspiracy in Bannack City and Virginia City. But this episode took place after Purple had left Montana, and so must have come from the later testimony of one of his friends at Bannack City.*

After the hanging of Henry Plummer, Ned Ray and Buck Stinson—the next day their bodies were taken from the gallows, and placed in an unoccupied two story building afterwards used by B. B. Burchett as a Hotel. This was the building George Chrisman occupied as a store on his first coming to Bannack. On the night after the removal of the remains, Dutch John (John Wagner) was taken to this place and hung to the beams of the second story. As he entered the building he turned to the body of Plummer lying upon the work bench and said, "That is the man who got me into all this trouble. Had I never seen him I might have lived an honest man." The weather was intensely cold, and the next day the bodies were frozen stark and stiff. Mozier was employed to bury them and taking a lumber wagon to where they lay, with the aid of an assistant, they were thrown into it, or rather slung in, like dead pigs. They were buried near the gallows upon which 3 of them expiated their dark and bloody crimes.

Why does'nt he write?

Nathaniel P. Langford dedicated his book, Vigilante Days and Ways: The Pioneers of the Rockies *(1890, vol. 1, p. v) with this illustration, which captures equally well the spirit of the closing paragraph of "The American Gold Miner."*

EPILOGUE

The American Gold Miner

The Arab with his camels, on his native desert, is not more migratory and nomadic in his habits than is the American Gold Miner in his. Wherever his lot is cast, he is the same restless character, ever seeking for Gold fields new, and placers rich; and this uneasy spirit of which he is possessed is not born of greed for filthy lucre, to hoard up with miserly care, for when found, he dispenses with liberal and prodigal old hand the favors which fortune has showered upon him. Usually not reckless, he is still indifferent to danger. In pursuit of the phantom that lures him on, he penetrates the recesses of the savage wilderness, bides the peltings of the Mountain storms, or toils and delves in the sweltering heat of the valleys below. With a benevolence of heart that blesses in substantial form his unfortunate suffering fellow, he adds that integrity of character which has made the name of "honest miner" a proverb in the land in which he dwells.

The simple laws ordained by himself and fellows, which define his rights and the tenure by which he holds his mining claim, are zealously upheld and respected; and woe be to him who attempts to violate their letter and spirit, for in their support he hesitates not an instant, to throw his Pistol and Bowie Knife into the wavering scale of right and Justice.

In sentiment he is an enthusiast. He regards his vocation as a mere pastime of his life, which is soon to end when he has made the rich strike, which is his constant daydream. Year after year still finds him hoping, toiling on, and not one in fifty ever reaches the goal of their cherished desires. Few indeed ever return to their early homes, the recollection of which is never lost, and always spoken of with affectionate tenderness.

It is remarkable how attached men from the same town, county or state, although strangers, will become to each other in his far off country. It offers the opportunity of conversing, perhaps for the first time in years, of home, maybe of parents and friends, the scenes of childhood and youth, of which amid all their wild wanderings they return an ardent and warm remembrance.

Alas! How many of these noble fellows have died in this pursuit of Gold. Their bones have whitened and lie bleaching on the ground by the lonely river's bank and in the distant mountain fastnesses, where they fell by the hand of the merciless Indian. Or the little hillocks of waving turf that line the highways and by-paths that led to these new territories, mark the final resting places of the pioneers of western progress and civilization.

INDEX

ABOUT THE AUTHOR

With a career-long interest in Montana's early political and legal history, Kenneth N. Owens has written extensively about territorial government and politics in the New Northwest. Among other works, Owens has edited an essay collection, *John Sutter and a Wider West* (1994), and *The Wreck of Sv. Nikolai* (1985), an account of the first Russian expedition to reach the Oregon Country. He is currently professor of history and director of the Capital Campus Public History Program at California State University, Sacramento.